Cotton in the kitchen

Wall Hanging ■ Instructions on page 2.

Wall Hanging, *shown on page 1.*

MATERIALS:

Unbleached Indian cotton fabric, 62cm by 94cm. Plaids in green shades, 82cm by 121cm. Fabric for appliqué: Sheeting: White, 41cm by 17cm; moss green, 27cm by 20cm; beige, 20cm square; gray, 30cm by 8cm; red, 23cm by 6cm; yellow, 14cm by 10cm; green, 14cm by 6cm; dark brown, 11cm by 10cm; ivory, 8cm by 10cm; ash brown, 6cm by 8cm; scraps of olive green and yellow green. Indian cotton fabric: Red, 26cm by 10cm; ash brown, 20cm by 6cm; light brown, 8cm by 4cm. Felt: Green and ocher, 20cm by 10cm each; light brown, 20cm by 5cm. Denim: Moss green, 20cm by 6cm and beige, 5cm by 8cm. Plaids: Red shades, 4cm square and gray shades, 5cm by 22cm.

CUTTING DIAGRAMS

Add 2cm for seam allowance all around unless otherwise indicated in parentheses.

Pieces for border plaids

Background
Unbleached Indian fabric
Cut 1.

Backing
Plaids   Cut 1.

90

58

90

a
Cut 2.
6

(4)
b
Cut 1.
64
(4)
←13→

(4)
c
Cut 1.
64
(4)
6

Cotton embroidery thread, No. 4: 2 skeins of dark brown. DMC six strand embroidery floss, No. 25: One

DIRECTIONS:

① Embroider and appliqué on unbleached Indian cotton.

② Place Indian cotton on plaids with wrong sides facing and bind edges with pieces for border in alphabetical order.

Backing (wrong side)

Background a

With right sides facing, sew border pieces onto background. Turn border to back and slip-stitch.

③ Back-stitch with one strand of dark brown, No. 4 catching backing fabric.

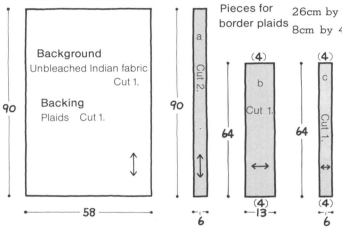

④ Twist 2 lengths of 170cm long cord and attach to dowel.

Wooden dowel
6cm
b
99 cm
a   a
c
64 cm
3cm
3cm

TOMATO SAUCE SPAGHETTI
INGRE-DIENTS   SPAGHETTI
30 cm
ONION   TOMATO GARLIC   SALT PEPPER   WINE

CHOP ONIONS AND GARLICS   HOW TO MAKE
MUSHED BOILED TOMATOES
30 cm

BOIL WATER
SERVE HOT!
30 cm

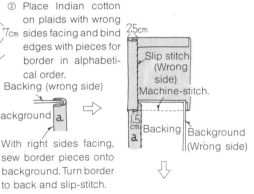

25cm
Slip stitch (Wrong side)
Machine-stitch.
1.5 cm   Backing   Background (Wrong side)

6 cm   7cm   b
a   Turn in seam allowance and slip-stitch
Back
[Sew piece (c) in same manner as fo

2

skein of umber (738); half skein each of beaver gray (647, 844), white, beige brown (841), almond green (503) and peacock green (255); small amount each of parakeet green (904), umber (434), smoke gray (640), ash gray (762) and colors matching with appliqué pieces.

Golden brown cotton thread. Wooden dowel, 3cm in diameter and 70cm long. Cotton cord, 340cm long. FINISHED SIZE: 64cm by 99cm.

Appliqué and Embroidery Patterns

Use sheeting for appliqué unless otherwise indicated.
Add no seam allowance to felt. Add 0.5cm for seam allowance all around to sheeting.
Slip-stitch appliqué pieces with 2 strands of embroidery floss No. 25 in matching color.
Embroider with 6 strands of floss in back stitch unless otherwise indicated.

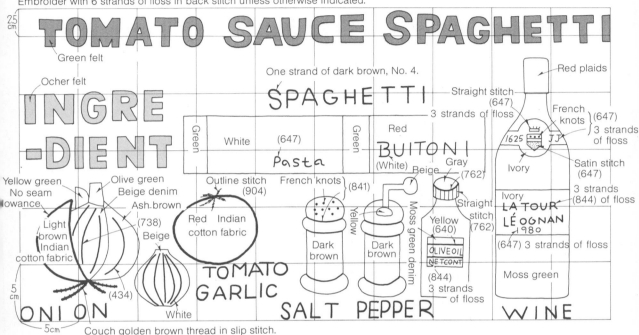

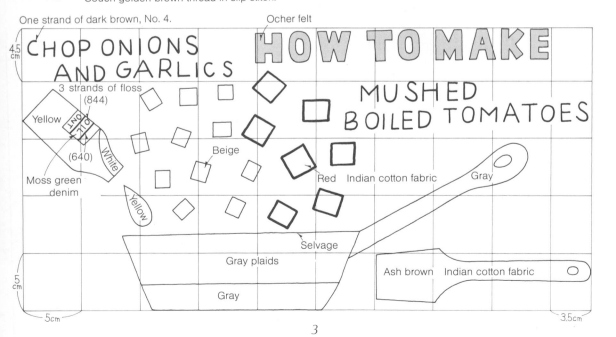

3

Checked Apron, Tea Cozy and Placemats ■ Instructions on page 6.

Checked Apron, Tea Cozy and Placemats, *shown on pages 4 & 5.*

Tea Cozy
MATERIALS:
Navy and white checks, 86cm by 32cm. Red sheeting for lining, 90cm by 24cm. Fabric for appliqué: Sheeting: Red, 9cm by 7cm; white, 6cm by 2cm; red dots, 5cm square. Thick quilt batting, 86cm by 32cm. Glue.
FINISHED SIZE: See diagrams.

CUTTING DIAGRAMS

Add 1cm for seam allowance unless otherwise indicated in parentheses.

(1.5)

Front and Back
Cut 2 each from checks and batting.

29

Fold line

4

36

(1.5)

Lining
Cut 2 from red sheeting.

21

35

Loop
Cut 1 each from red sheeting and batting.

(1.5)

15

(1.5) 3

Top
7.5

Cut 1 each from checks, red sheeting and batting.

Actual-size Patterns

Glue appliqué pieces onto background and zigzag stitch by machine.

DIRECTIONS:

④ Place red sheeting on batting, roll and slip-stitch.

Batting

Slip-stitch.

1.5 cm

⑤ Gather top edge of body. Sew top pieces onto body catching loop in between. Slip-stitch lining of top piece onto wrong side.

Top Slip-stitch.
Loop
Batting
Outer p
Lining of top / Slip-stitch. Linin
Gather.

① Appliqué onto front.

25cm

TEA

4 cm

36cm

② Place checked piece on batting and baste. Sew front and back pieces together with right sides facing. Turn to right side.

⑥ Turn bottom edge to back and slip-stitch lining.

Batting | Lining | Slip stitch
Front piece | 4cm

③ Sew front and ba pieces for lining together in same mar Insert lining into outer p

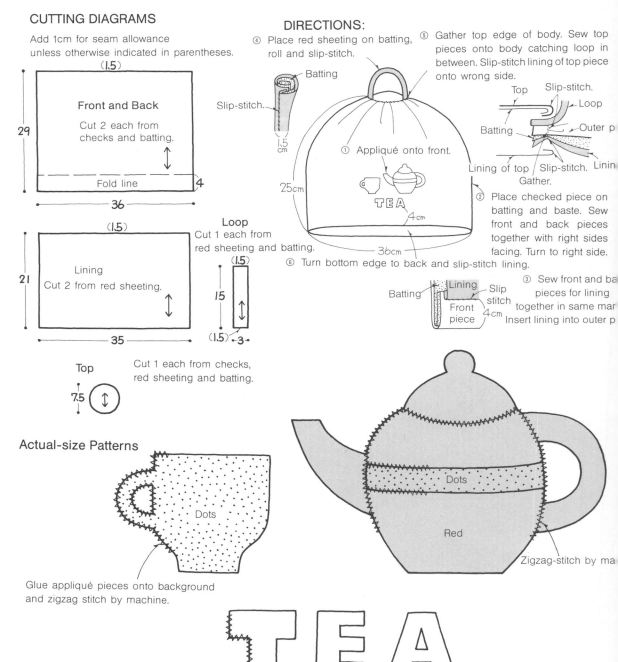

Dots

Dots

Red

Zigzag-stitch by ma

TEA

6

Apron

MATERIALS:

Navy checked cotton fabric, 90cm by 2m.
Fabric for appliqué: Red sheeting, 9cm by
7cm; red dots, 5cm square. Six red
buttons, 1.3cm in diameter. Glue.

FINISHED SIZE: Bust, 104.5cm. Length,
88cm.

DIRECTIONS:

① Sew shoulder seams.
Turn seams to one side and
top-stitch. (See page 85.)

⑤ Sew facing onto body.

⑦ Make buttonholes.
Sew on buttons.

Body (Right side)

Machine

Facing (Right side)

0.3cm

Top-stitch

Facing (Wrong side)

Body (Wrong side)

6.5cm

⑧ Appliqué in place.

⑥ Machine-stitch top of pocket and sew onto body.

3cm

0.3cm

③ Fold bottom edge twice and machine-stitch.

(Wrong side)

3.5cm

④ Fold facing and machine-stitch.

② Sew side seams. Turn seams to one side and top-stitch.

CUTTING DIAGRAMS

Add 1cm for seam allowance unless otherwise indicated in parentheses.
Make buttonholes in left back piece.

Cut facing of front and back into one piece.

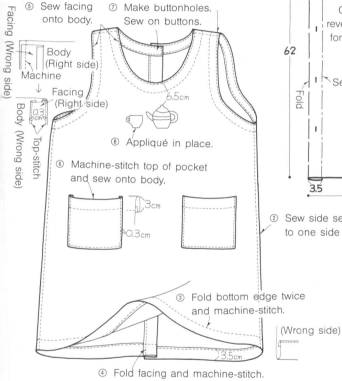

Back
Cut 2 pieces,
reversing pattern
for opposite side.

Facing

Selvage

Fold

10 7   7 10
5   3 3   3 3 3.5
3   9
12.5   3
21   15
2 1.5   11
27 Facing 27
6 10   10 6
3   3
62   62
      Front
      Cut 1
      (4)
      Pocket
      Cut 2
6   Fold
(1.5) (1.5)
14
14
28
0.5
3.5   (4.5) (4.5)
33   33

Placemat

MATERIALS FOR ONE:

Navy and white checks, 67cm by 45cm. Navy checked
cotton fabric, 27cm by 22cm. Fabric for appliqué: Red
dots for cup, 5cm square; for pot, red sheeting, 9cm by
7cm and red dots, 5cm by 1cm. Glue.

FINISHED SIZE: 43cm by 32.5cm.

CUTTING DIAGRAM AND DIRECTIONS:

Add 1cm for seam allowance all around.

③ With right sides facing,
sew pieces together leaving
10cm open for turning.
Turn inside out
and machine-stitch.

Opening for turning

Small checks

Large checks

① Appliqué cup or pot in place.

② Sew appliquéd piece onto ground fabric.

10
32.5   19.5
25   6.5
9
Fold
43

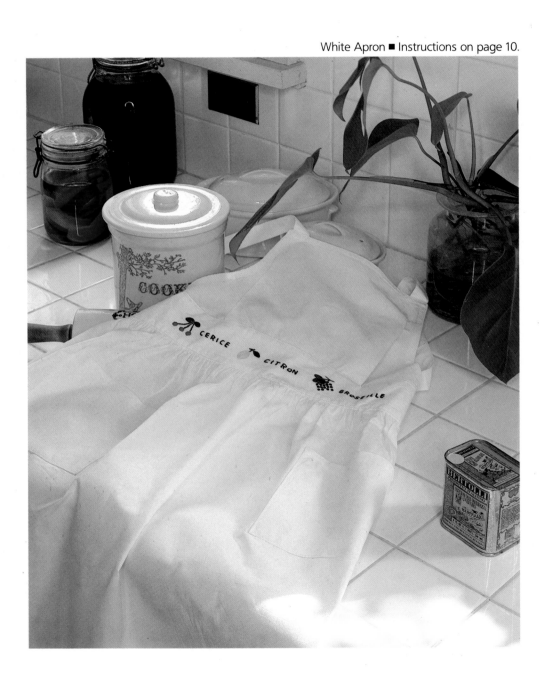

White Apron and Mitts,
shown on pages 8 & 9.

Apron

MATERIALS:

White sheeting, 90cm by 180cm. Six strand embroidery floss, No. 25: Half skein of charcoal gray; small amount each of moss green, olive green, turquoise, red, lemon yellow and blue. Four buttons, 2cm in diameter.

FINISHED SIZE: See diagram.

CUTTING DIAGRAMS
Add seam allowance indicated in parentheses.

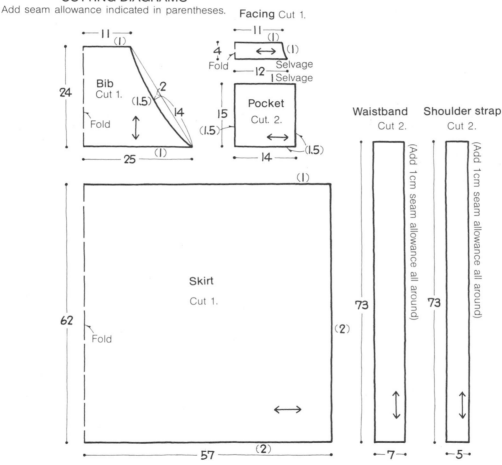

Facing Cut 1.

Bib Cut 1.

Pocket Cut. 2.

Skirt Cut 1.

Waistband Cut 2.

Shoulder strap Cut 2.

Actual-size Embroidery Pattern

Use 3 strands of floss and outline stitch unless otherwise indicated.

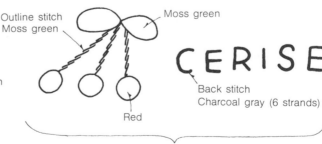

Outline stitch
Moss green

Moss green

Red

CERISE

Back stitch
Charcoal gray (6 strands)

For Mitt A

DIRECTIONS

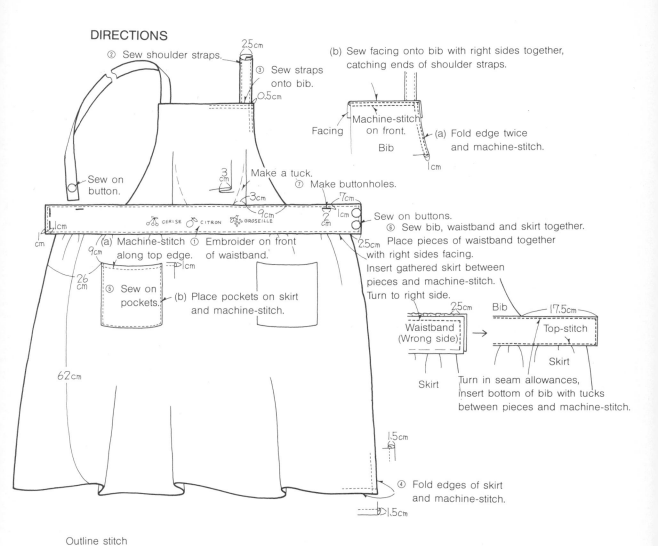

② Sew shoulder straps.

25cm

③ Sew straps onto bib.

0.5cm

(b) Sew facing onto bib with right sides together, catching ends of shoulder straps.

Facing

Machine-stitch on front.

Bib

(a) Fold edge twice and machine-stitch.

1cm

Sew on button.

3cm

Make a tuck.

3cm

9cm

⑦ Make buttonholes.

7cm

2cm

1cm

Sew on buttons.

⑥ Sew bib, waistband and skirt together.

Place pieces of waistband together with right sides facing. Insert gathered skirt between pieces and machine-stitch. Turn to right side.

1cm

cm

(a) Machine-stitch ① Embroider on front along top edge. of waistband.

1cm

9cm

⑤ Sew on pockets.

(b) Place pockets on skirt and machine-stitch.

CERISE   CITRON   GROSEILLE

25cm

26cm

62cm

25cm

Waistband (Wrong side)

Bib

17.5cm

Top-stitch

Skirt

Skirt

Turn in seam allowances, insert bottom of bib with tucks between pieces and machine-stitch.

1.5cm

④ Fold edges of skirt and machine-stitch.

1.5cm

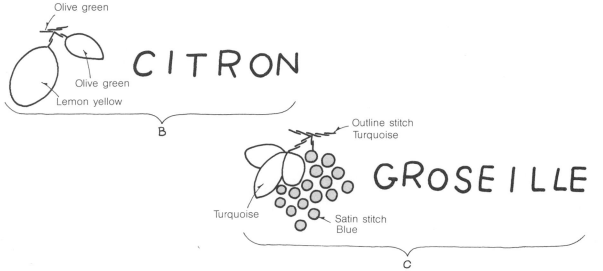

Outline stitch
Olive green

Olive green

Lemon yellow

CITRON

B

Outline stitch
Turquoise

Turquoise

Satin stitch
Blue

GROSEILLE

C

See page 69 for Hot Mitts.

Kitchen towels ■ Instructions on page 15.

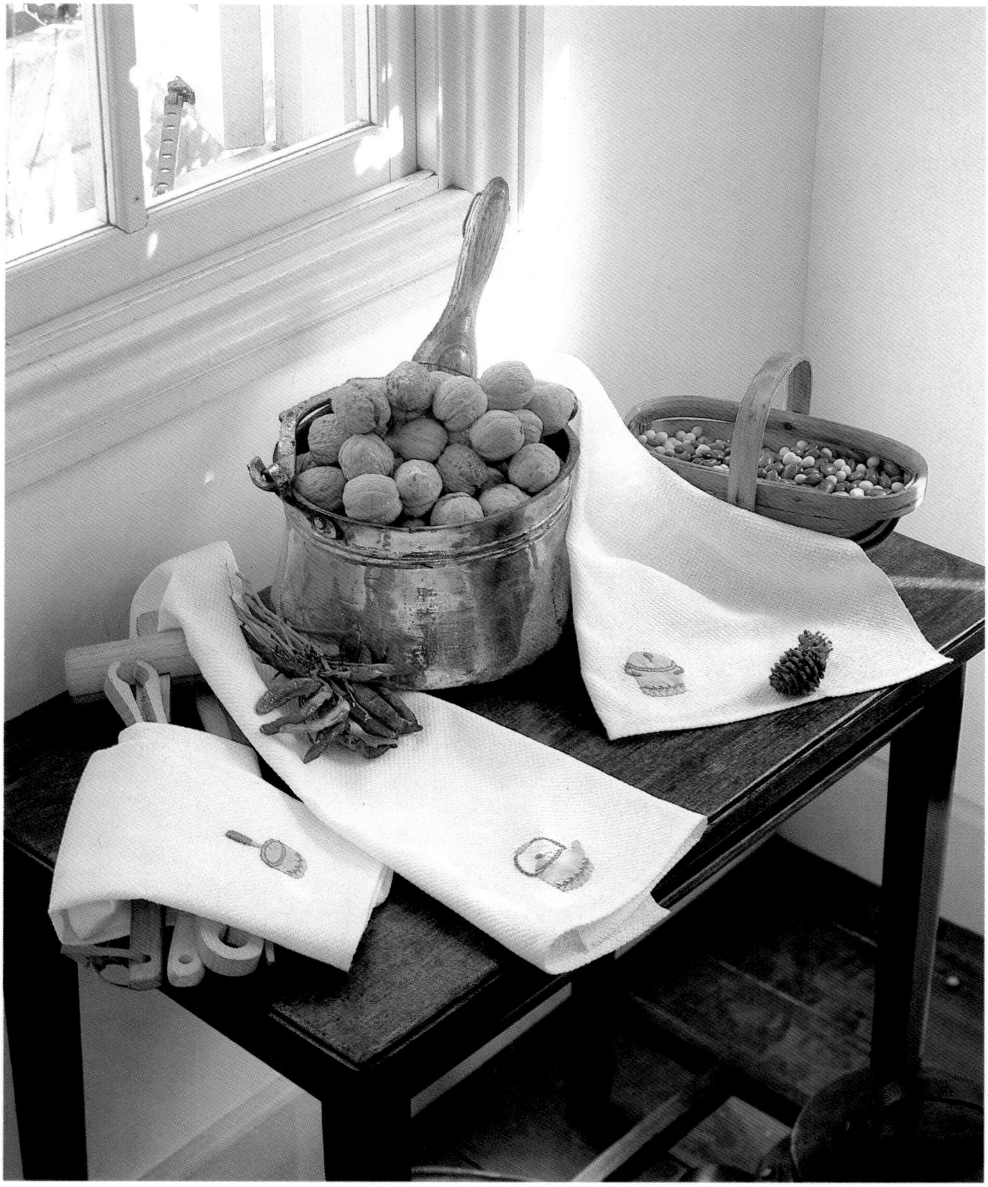

Cutlery Case, *shown on page 12.*

MATERIALS:
White flannel, 76cm by 45cm. Six strand embroidery floss, No. 25: One skein of silver gray. Gray binding tape, 165cm long. One button, 1.2cm in diameter. Thin quilt batting, 38cm by 26cm.
FINISHED SIZE: See diagram.

CUTTING DIAGRAMS
Add 1cm for seam allowance all around.

Actual-size Embroidery Pattern
Use 6 strands of floss in silver gray and back stitch.

Top and Inside
Cut 2 from flannel and 1 from batting.

24

36

Pocket
Cut 2 from flannel.

10

36

Flap Cut 2 from flannel.

5

⑥ Baste batting onto wrong side of inner piece.
With right sides facing, sew outer and inner pieces together, catching binding tape and loop and leaving 11cm open for turning. Turn to right side and slip-stitch opening closed.

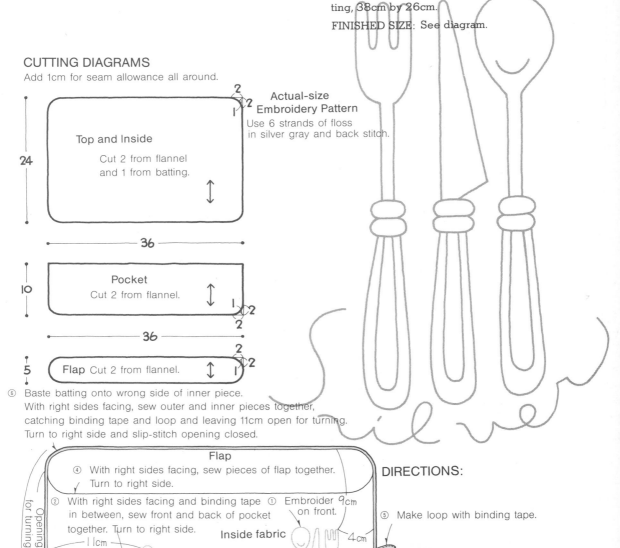

Flap
④ With right sides facing, sew pieces of flap together. Turn to right side.

② With right sides facing and binding tape in between, sew front and back of pocket together. Turn to right side.

① Embroider on front.

Inside fabric

DIRECTIONS:

⑤ Make loop with binding tape.

9cm

4cm

5cm   7cm   0.5cm

Opening for turning

24cm

11cm

⑦ Sew on button.

6cm
(Wrong side)
(Right side)

Binding tape

12.5cm

Pocket

③ Place pocket on inner piece and stitch at 6cm intervals.

36cm

Kichen Towels, *shown on page 13.*

MATERIALS FOR ONE:

Kitchen towel, 35cm by 44cm. Fabric for appliqué: Blue cotton fabric, 5cm square. Six strand embroidery floss, No. 25 in blue gray.

FINISHED SIZE: 35cm by 44cm.

Place for appliqué

3cm

35cm

Actual-size Appliqué Patterns

Add 0.3cm for seam allowance.
Use 3 strands of floss.

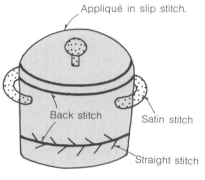

Appliqué in slip stitch.

Back stitch

Satin stitch

Straight stitch

Continued from page 3.

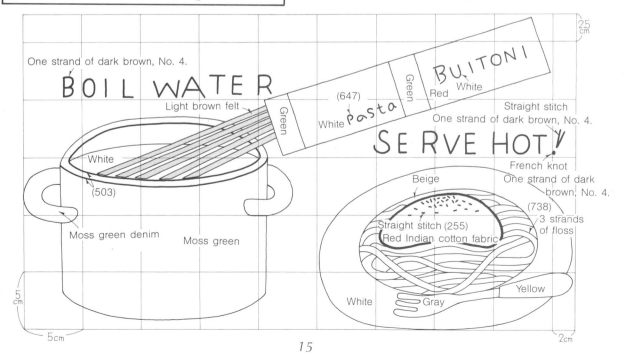

One strand of dark brown, No. 4.

BOIL WATER

Light brown felt

Green

White Pasta

(647)

Green

Red White

BUITONI

Straight stitch
One strand of dark brown, No. 4.

SE RVE HOT!

French knot
One strand of dark brown, No. 4.

White

(503)

Moss green denim

Moss green

Beige

Straight stitch (255)
Red Indian cotton fabric

(738)

3 strands of floss

White

Gray

Yellow

5 cm

5cm

2cm

25 cm

15

Mat ■ Instructions on page 18.

Patched Tote Bag ■ Instructions on page 67.

Mat, *shown on page 16.*

MATERIALS:

Brown with white flowers, 86cm by 70cm. Brown checks, 82cm by 48cm. Green sheeting, 50cm by 66cm. Beige bias tape, 2cm wide by 201cm long. Six strand embroidery floss, No. 25: Two skeins of off white; half skein each of brown and green. Thick quilt batting, 85cm by 66cm.

FINISHED SIZE: 62cm by 47cm.

CUTTING DIAGRAMS

Add 2cm for seam allowance unless otherwise indicated in parentheses.

Fabric for patchwork

Floral print

Checks

Cut 3.   Cut 8.   Cut 2.   Cut 6.

a	b	c	d
31	31	44	44
8	25	8	25
(l)	(l)	(l)	(l)

Background
Cut 1 each from checks and batting.
44
31

e Beige bias tape   Cut 3.
2⁝
35
No seam allowance

f   Cut 2.
2⁝
48

Border
Green sheeting
Quilt batting

A	B
Cut 2.	Cut 2.
62	31
8	9

Back
Floral print
Cut 1.
62
47

Actual-size Embroidery Patterns

Use 6 strands of floss.

Running stitch in brown

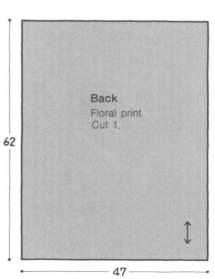

Back stitch in green

KITCHEN

DIRECTIONS

① Place piece (e) on (a) and baste.

③ Place pieces on background fabric from left as shown in the diagram and baste.

Seam allowance

Background

④ Weave pieces starting from top and baste.

⑥ Quilt along quilting lines with 3 strands of white.

⑤ Embroider letters

KITCHEN

② Place piece (f) on (c) and baste.

Opening for turning

Embroider tea pot on right side.

KITCHEN

KITCHEN

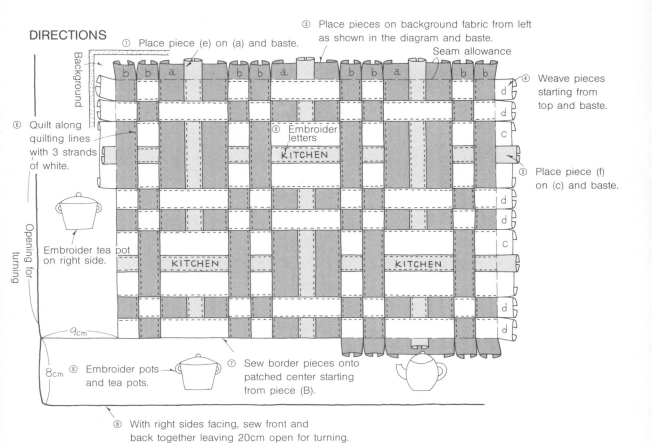

9cm

8cm ⑨ Embroider pots and tea pots.

⑦ Sew border pieces onto patched center starting from piece (B).

⑧ With right sides facing, sew front and back together leaving 20cm open for turning. Turn to right side and slip-slitch opening closed.

Cotton
in my room

Comforter and Pillow Case ■ Instructions on page 22.

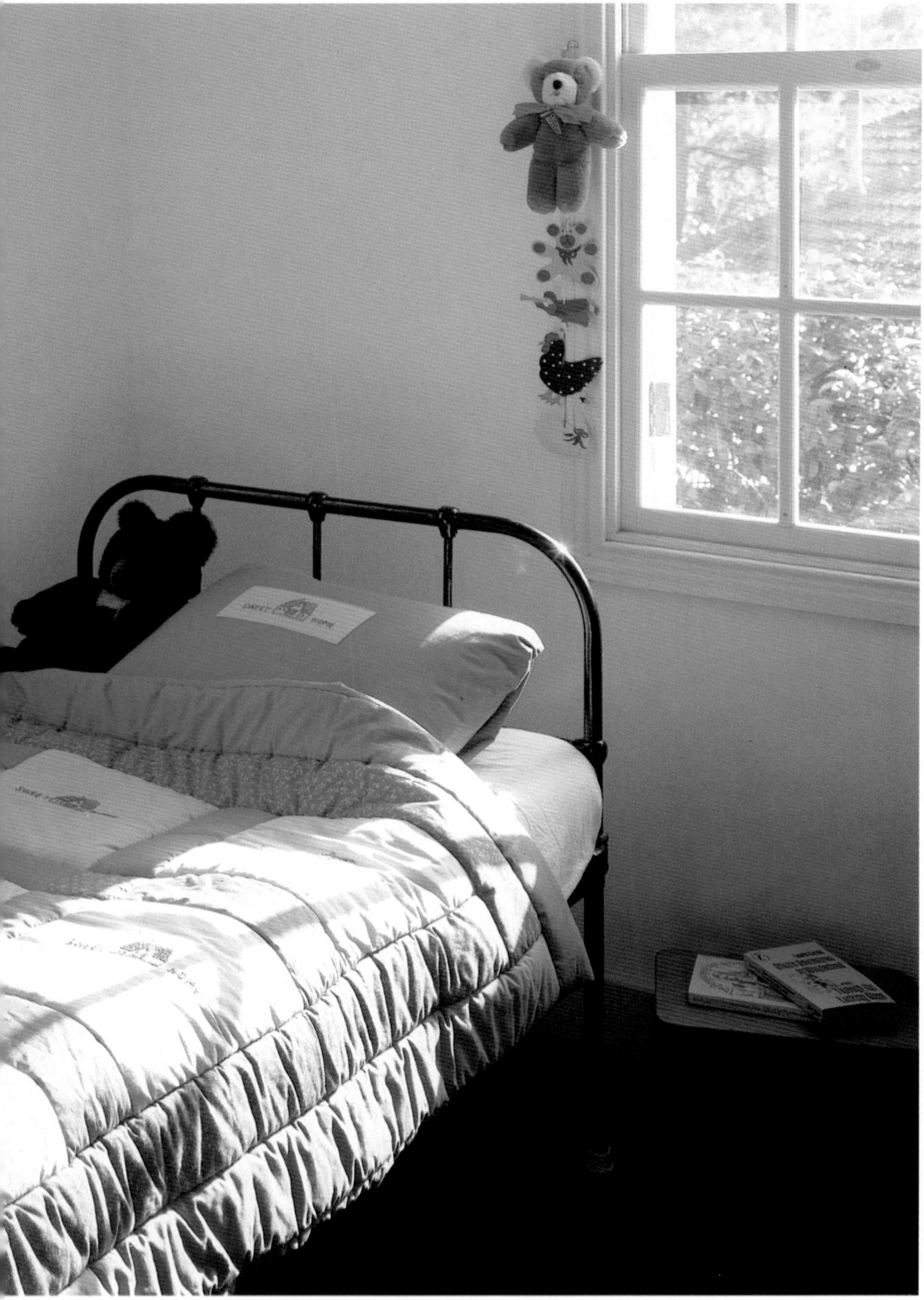

Comforter and Pillow Case,
shown on pages 20 & 21.

Comforter
MATERIALS:
Cotton broadcloth: Blue, 90cm by 540cm; light blue, 66cm by 2m; print, 65cm by 155cm. Thick quilt batting, 90cm by 480cm. Six strand embroidery floss, No. 25: 6 skeins of blue gray.
FINISHED SIZE: 150cm by 190cm.

CUTTING DIAGRAMS
Add 1cm for seam allowance unless otherwise indicated in parentheses.

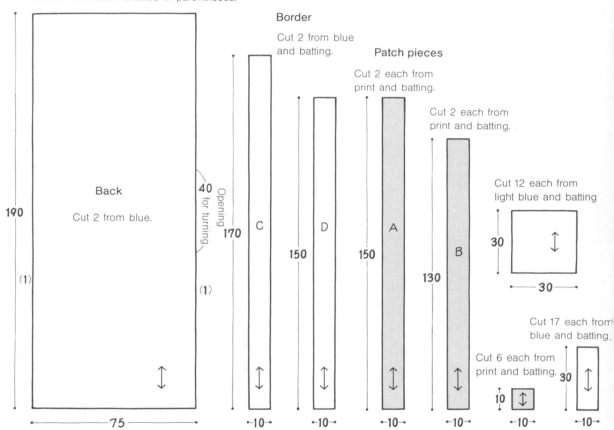

Back
Cut 2 from blue.

190
(1)
75

40
Opening for turning
170
(1)

Border
Cut 2 from blue and batting.

C
150
10

D
150
10

Patch pieces
Cut 2 each from print and batting.

A
150
10

Cut 2 each from print and batting.

B
130
10

Cut 12 each from light blue and batting
30
30

Cut 17 each from blue and batting.
30
10

Cut 6 each from print and batting.
10
10

Actual-size Embroidery Pattern

Use 6 strands of floss in blue gray and back stitch unless otherwise indicated.

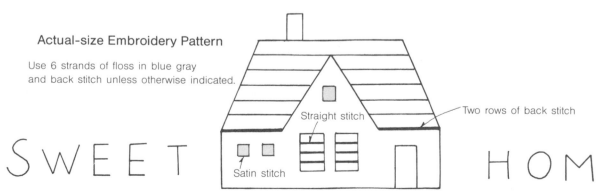

Straight stitch

Two rows of back stitch

Satin stitch

SWEET   HOME

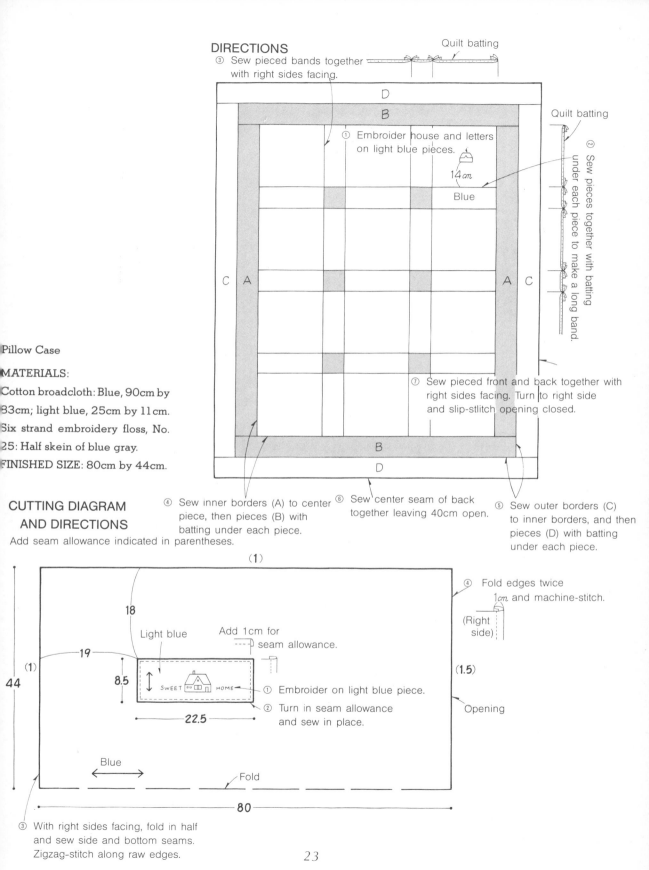

DIRECTIONS

③ Sew pieced bands together with right sides facing.

Quilt batting

D

B

① Embroider house and letters on light blue pieces.

14 cm

Blue

Quilt batting

② Sew pieces together with batting under each piece to make a long band.

C A

A C

D

B

D

⑦ Sew pieced front and back together with right sides facing. Turn to right side and slip-stitch opening closed.

④ Sew inner borders (A) to center piece, then pieces (B) with batting under each piece.

⑥ Sew center seam of back together leaving 40cm open.

⑤ Sew outer borders (C) to inner borders, and then pieces (D) with batting under each piece.

Pillow Case

MATERIALS:

Cotton broadcloth: Blue, 90cm by 83cm; light blue, 25cm by 11cm.

Six strand embroidery floss, No. 25: Half skein of blue gray.

FINISHED SIZE: 80cm by 44cm.

CUTTING DIAGRAM AND DIRECTIONS

Add seam allowance indicated in parentheses.

（1）

④ Fold edges twice 1cm and machine-stitch.

(Right side)

18

Light blue

Add 1cm for seam allowance.

19

(1)

8.5

SWEET HOME

① Embroider on light blue piece.

② Turn in seam allowance and sew in place.

(1.5)

Opening

44

22.5

Blue

Fold

80

③ With right sides facing, fold in half and sew side and bottom seams. Zigzag-stitch along raw edges.

23

Sewing Box, Pincushion and Scissors Case Set ■ Instructions on page 26.

Camisole and Panties ■ Instructions on page 34.

Sewing Box, Pincushion and Scissors Case Set, *shown on page 24.*

Sewing Box

MATERIALS:

Floral print: Red background, 40cm square; white background, 26cm by 18cm. Ready-made basket, 18cm by 12.5cm oval and 7cm deep. Cardboard, 24cm by 18cm. Red ribbon, 1cm by 79cm. Thick quilt batting, 18cm by 12.5cm. Glue.

FINISHED SIZE: Same size as basket.

CUTTING DIAGRAMS

Add 0.5cm for seam allowance all around.

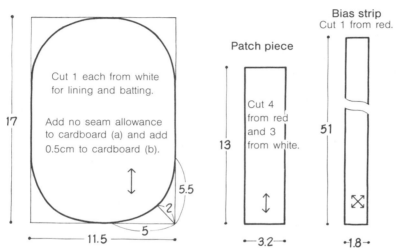

Cut 1 each from white for lining and batting.

Add no seam allowance to cardboard (a) and add 0.5cm to cardboard (b).

17

11.5

5.5

2

5

Patch piece

Cut 4 from red and 3 from white.

13

3.2

Bias strip
Cut 1 from red.

51

1.8

DIRECTIONS:

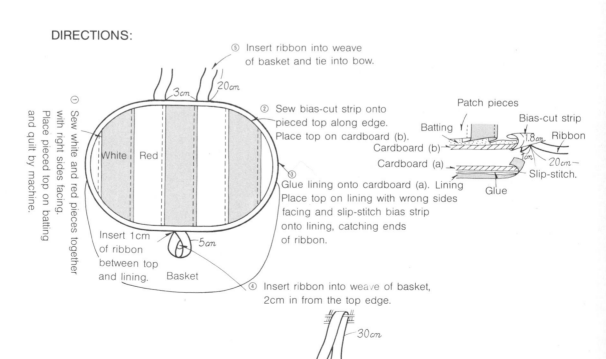

① Sew white and red pieces together with right sides facing. Place pieced top on batting and quilt by machine.

White Red

Insert 1cm of ribbon between top and lining.

Basket

3cm

20cm

5cm

⑤ Insert ribbon into weave of basket and tie into bow.

② Sew bias-cut strip onto pieced top along edge. Place top on cardboard (b).

③ Glue lining onto cardboard (a). Lining Place top on lining with wrong sides facing and slip-stitch bias strip onto lining, catching ends of ribbon.

④ Insert ribbon into weave of basket, 2cm in from the top edge.

Patch pieces

Batting

Cardboard (b)

Cardboard (a)

1.8cm

Bias-cut strip

Ribbon

20cm

Slip-stitch.

Glue

30cm

Pincushion

MATERIALS:

Floral print: Red background, 22cm by 9cm; white background, 10cm by 9cm. Red ribbon, 1cm by 46cm. Small amount of polyester fiberfill.

FINISHED SIZE: 11cm by 8cm.

CUTTING DIAGRAM

Add 0.5cm for seam allowance all around.

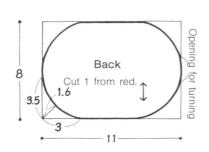

8
3.5
1.6
3
11
Back
Cut 1 from red.
Opening for turning

Patch piece

Cut 4 each from red and white.

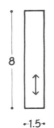

8
-1.5-

DIRECTIONS

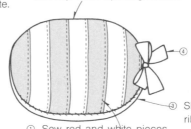

② With right sides facing, sew front and back together leaving side open for turning. Turn to right side. Stuff with fiberfill and slip-stitch opening closed.

④ Tie 17cm long ribbon into bow and sew in place.

③ Slip-stitch 29cm long ribbon around side.

① Sew red and white pieces together with right sides facing. Turn seams to one side and top-stitch on front along seams.

Scissors Case

MATERIALS:

Floral print: White background, 34cm by 25cm; red background, 16cm by 6cm. Thin quilt batting, 12cm by 11cm.

FINISHED SIZE: 6cm by 11cm.

CUTTING DIAGRAMS

Add 0.5cm for seam allowance all around.

Patch piece
Cut 4 each from red and white.

5
-1.3-

Border
Cut 2 from red.

5
-2-

Bias strip
Cut 1 from white.

30
-1.8-

Cut 1 for Back and 2 for lining from white, and 2 from batting.

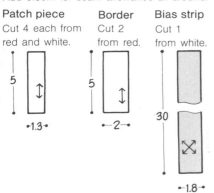

10
5
6
1.9

① Sew red and white pieces together with right sides facing. Place pieced top on batting and quilt along seams by machine.

DIRECTIONS

③ With wrong sides facing, bind edges with bias strip. And make the loop with remaining bias strip.

Slip-stitch.

Red
White

② Place quilted top on lining and bind top edges with bias-cut strip (see page 84).

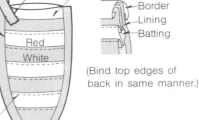

Border
Lining
Batting

(Bind top edges of back in same manner.)

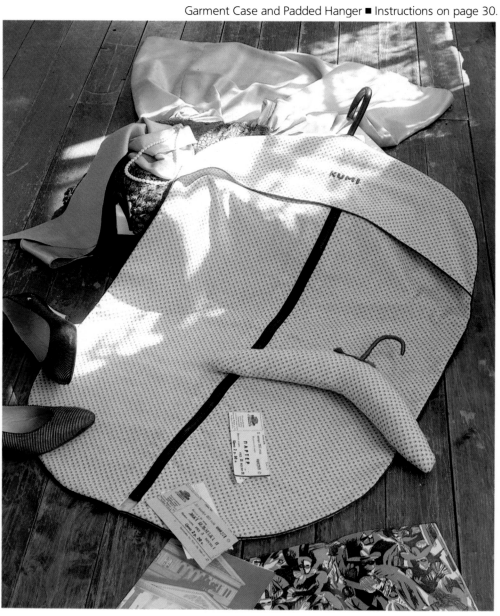

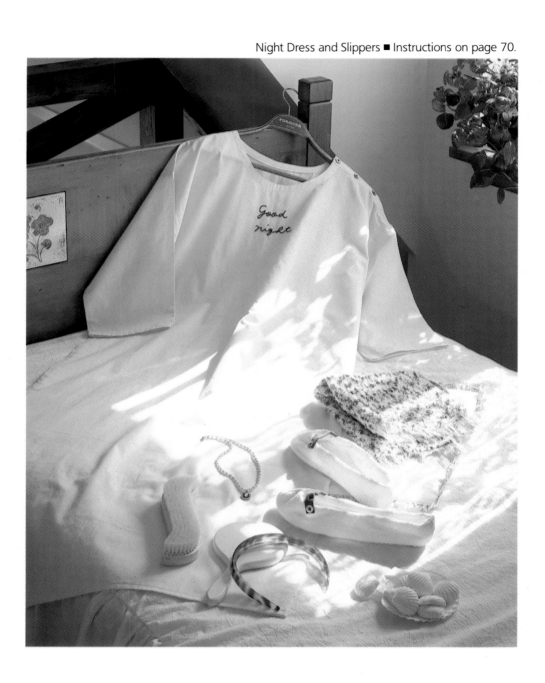

Garment Case and Padded Hanger, *shown on page 28.*

Garment Case

MATERIALS:

Floral print and light blue denim, 90cm by 182cm each.

Six strand embroidery floss, No. 25: Half skein of blue.

Navy open zipper, 85cm long. Navy bias tape, 3.3m long.

FINISHED SIZE: 56cm by 44cm.

CUTTING DIAGRAMS

Add 1cm for seam allowance unless otherwise indicated in parentheses.

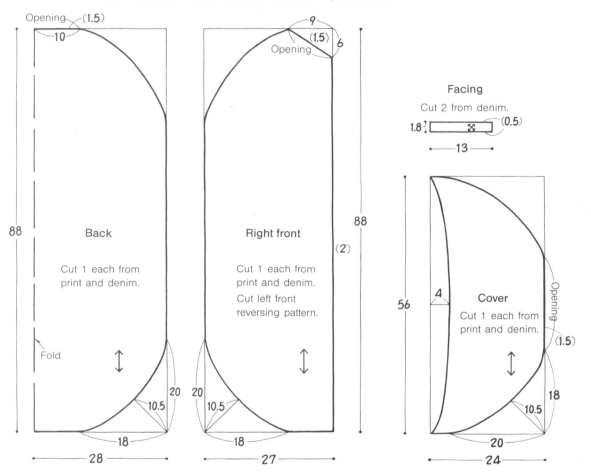

Opening (1.5)
10

Back

Cut 1 each from print and denim.

Fold

88

20
10.5
18
28

9
(1.5)
Opening
6

Right front

Cut 1 each from print and denim.
Cut left front reversing pattern.

88
(2)

20 20
10.5 10.5
18 18
27

Facing

Cut 2 from denim.

1.8
13
(0.5)

Cover

Cut 1 each from print and denim.

56
4
Opening
(1.5)
18
10.5
20
24

Actual-size Embroidery Pattern

Use 6 strands of floss.

KUMI

Embroider your name.

Satin stitch.

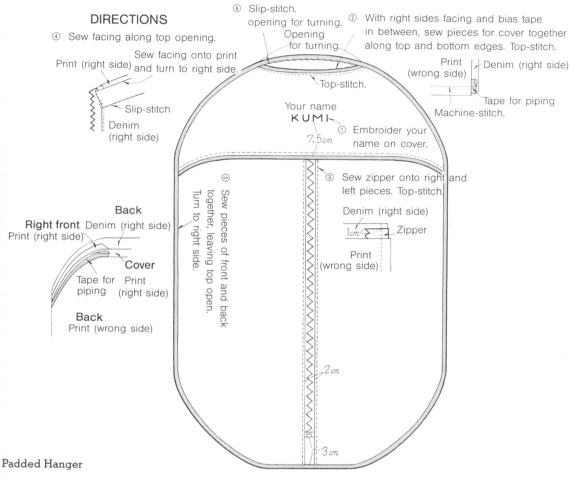

DIRECTIONS

④ Sew facing along top opening.

Sew facing onto print and turn to right side.

Print (right side)

Slip-stitch

Denim (right side)

⑥ Slip-stitch. opening for turning.
Opening for turning

Top-stitch.

② With right sides facing and bias tape in between, sew pieces for cover together along top and bottom edges. Top-stitch.

Print (wrong side) Denim (right side)

Tape for piping Machine-stitch.

Your name
KUMI
7.5cm

① Embroider your name on cover.

⑤ Sew pieces of front and back together, leaving top open. Turn to right side.

③ Sew zipper onto right and left pieces. Top-stitch.

Denim (right side)

1cm Zipper

Print (wrong side)

2cm

3cm

Back
Right front Denim (right side)
Print (right side)

Cover
Tape for piping Print (right side)

Back
Print (wrong side)

Padded Hanger

MATERIALS:

Floral print, 34cm by 42cm. Wire hanger. Blue ribbon, 1.2cm wide and 1m long. Polyester fiberfill, 60g. Glue.

FINISHED SIZE: See diagram.

CUTTING DIAGRAM

Add 1cm for seam allowance all around.

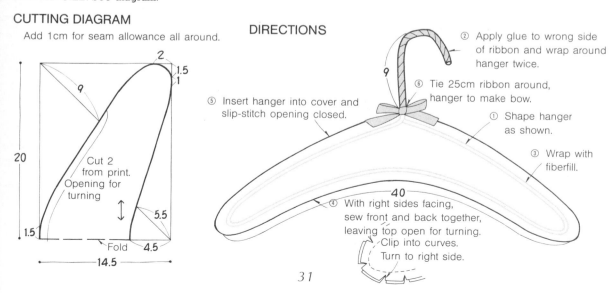

2
1.5
1
9

20

Cut 2 from print.
Opening for turning

5.5

1.5
Fold 4.5

14.5

DIRECTIONS

② Apply glue to wrong side of ribbon and wrap around hanger twice.

9

⑥ Tie 25cm ribbon around, hanger to make bow.

① Shape hanger as shown.

③ Wrap with fiberfill.

⑤ Insert hanger into cover and slip-stitch opening closed.

40

④ With right sides facing, sew front and back together, leaving top open for turning. Clip into curves. Turn to right side.

31

Pillows, *shown on pages 32 & 33.*

MATERIALS FOR ONE:
Quilted fabric, 75cm by 33cm. Make one pair each of red, navy, unbleached, red and white checks (large and small), navy and white checks (large), red and green checks, red and green plaids; one each of navy and white checks (small) and prints. Inner pillow, 30cm square.
FINISHED SIZE: 30cm square.

CUTTING DIAGRAM AND DIRECTIONS
Add seam allowance indicated in parentheses.

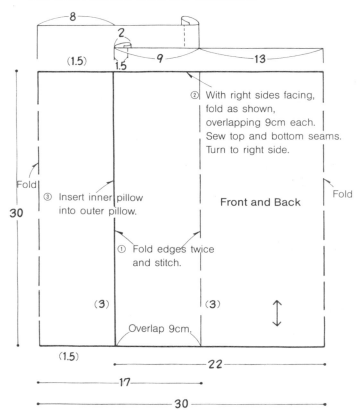

Camisole and Panties,
shown on page 25.

MATERIALS:
Floral cotton jersey: 50cm by 120cm for camisole; 82cm by 35cm for panties; 42cm by 28cm for case. Elastic tape, 0.3cm by 74cm. Cotton cord, 65cm long.
FINISHED SIZE: Camisole: Bust, 88cm. Length, 55cm. See diagrams for sizes of panties and case.

CUTTING DIAGRAMS

Add 1.5cm for seam allowance all around unless otherwise indicated in parentheses.

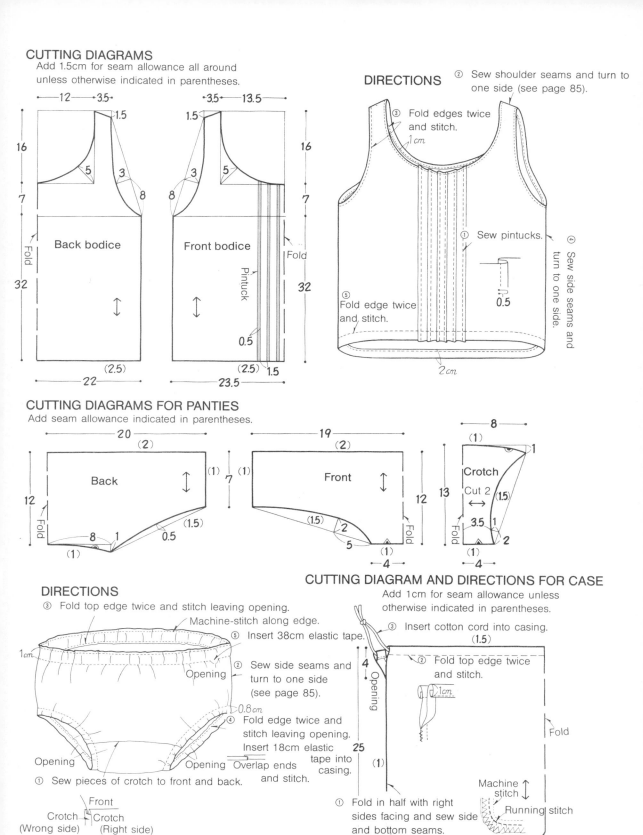

DIRECTIONS

② Sew shoulder seams and turn to one side (see page 85).

③ Fold edges twice and stitch. 1cm

① Sew pintucks. 0.5

④ Sew side seams and turn to one side.

⑤ Fold edge twice and stitch.

2cm

Back bodice

Front bodice

Pintuck 0.5

Fold

12 · 3.5 ·   · 3.5 · 13.5

16   16
7   7
32   32

1.5   1.5
5   5
3   3
8   8

(2.5)   (2.5) 1.5
22   23.5

CUTTING DIAGRAMS FOR PANTIES

Add seam allowance indicated in parentheses.

Back (20) (2)   Front (19) (2)   Crotch Cut 2 (8) (1)

12   7   12   13
Fold   Fold   Fold

8 1 (1.5) 0.5   (1.5) 2 5 (1)   (1.5) 1 3.5 1 2
(1)   4   (1) 4

DIRECTIONS

③ Fold top edge twice and stitch leaving opening. Machine-stitch along edge.

⑤ Insert 38cm elastic tape.

② Sew side seams and turn to one side (see page 85).

④ Fold edge twice and stitch leaving opening. Insert 18cm elastic tape into casing. Overlap ends and stitch.

① Sew pieces of crotch to front and back.

1cm   Opening   0.8cm   Opening   Opening   Opening

Front
Crotch   Crotch
(Wrong side)   (Right side)

CUTTING DIAGRAM AND DIRECTIONS FOR CASE

Add 1cm for seam allowance unless otherwise indicated in parentheses.

③ Insert cotton cord into casing. (1.5)

② Fold top edge twice and stitch. 1cm

4   Opening   25   (1)   Fold

Machine stitch   Running stitch

① Fold in half with right sides facing and sew side and bottom seams.

(1)   20

Heartwarming Gifts

Gift Bag, Box and Tags ■ Instructions on page 38.

Gift Bag, Box and Tags,
shown on page 36.

Gift Bag

MATERIALS:

Quilted fabric: Unbleached, 32cm by 25cm; red, 32cm by 23cm; red and white checks, 32cm by 18cm. Red felt, 27cm by 2cm. Six strand embroidery floss, No. 25: Half skein of blue. Red ribbon, 2.5cm wide and 80cm long.

FINISHED SIZE: See diagram.

CUTTING DIAGRAMS

Add 1cm for seam allowance
unless otherwise indicated in parentheses.

Body (3)

7 | Cut 1 from red. ↕
30

23 | Cut 1 from unbleached. ↕
30

11 | Cut 1 from checks. ↕
30

Base

9.5 ↕

Cut 1 from red.

Handle

(2) Cut 1 from red felt.
2 ↕
23
No seam allowance

(2) Cut 1 from checks.
2.5 ↕
23

DIRECTIONS

⑦ Make handle and sew onto body.

(a) Zigzag-stitch.
2.5cm Checks
Felt
2cm
(b) Machine-stitch.

② Sew pieces together and turn seam to one side (see page 85).

① Embroider in place.

④ Sew side seam and turn seam to one side.
⑤ Fold top edge twice and zigzag-stitch.
2cm
Body (Wrong side)
1.5cm

⑧ Sew ribbon at center back and trim ends of ribbon.
Center of back
Red
40cm
Unbleached
37cm
3cm

Red
Unbleached

Best wine & love
2cm

Checks

Red

41cm

③ Sew pieces together and turn seam to one side. Top-stitch.

⑥ Sew base onto body with right sides facing.
Stitch
Base
Body

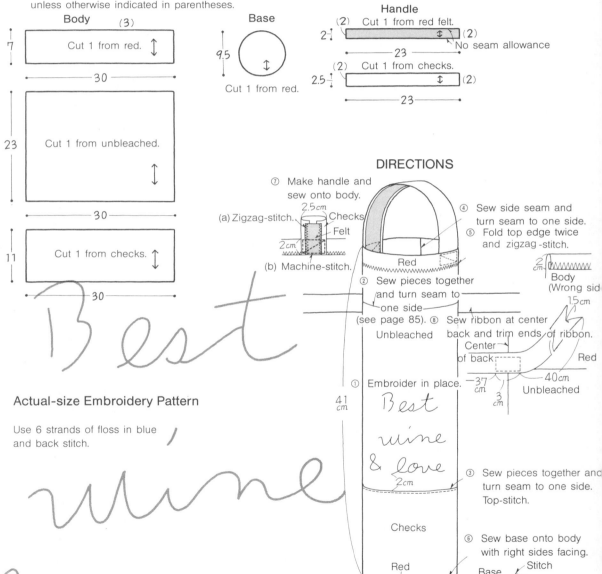

Actual-size Embroidery Pattern

Use 6 strands of floss in blue
and back stitch.

Best wine & love

CUTTING DIAGRAMS

Add 0.5cm for seam allowance unless otherwise indicated.

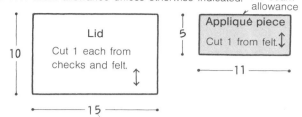

Lid
Cut 1 each from checks and felt.

10
15

No seam allowance

Appliqué piece
Cut 1 from felt.

5
11

5
10
5

Fold line

Box
Cut 1 each from red and felt.

5 — 15 — 5

Tag

MATERIALS FOR ONE:

Red felt, 14cm by 6cm. White cord, 22cm long. Six strand embroidery floss, No. 25 in white for tags A, C and D, and blue for B. Six pearl beads, 0.3cm in diameter for A. Star-shaped spangles, 1.5cm and 1cm in diameter, one each. Small beads, one transparent and one pink for B. Golden tape, 0.6cm by 7cm for C.
FINISHED SIZE: See diagram

Gift Box

MATERIALS:

Quilted fabric: Red, 26cm by 21cm; red and white checks, 16cm by 11cm. Red felt, 27cm by 32cm. Six strand embroidery floss, No. 25 in red and white. White cord, 6cm. One button, 1.3cm in diameter.
FINISHED SIZE: See diagram.

DIRECTIONS

① Embroider on felt and sew on checks.

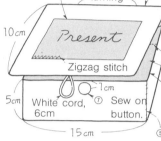

⑥ Catch stitch lid onto body using 6 strands of red floss.

Opening for turning

10cm

Present

Zigzag stitch

②

④ Machine-stitch along fold line.
Felt
③ Sew pieces for box together with right sides facing, leaving opening. Turn to right side and slip-stitch opening closed.

5cm
White cord, 6cm
1cm
⑦ Sew on button.

15cm

⑤ Overcast corners from right side.

② Sew pieces for lid together with right sides facing, leaving opening and catching ends of loop. Turn to right side and slip-stitch opening closed.

Actual-size Pattern

Use 6 strands of white floss and back stitch.

Present

Actual-size Embroidery Pattern

A
Use 6 strands of white floss and back stitch.

Cord

DIRECTIONS

Cut 2 from red felt adding no seam allowance.

① Embroider on front.

② Sew on beads.

③ Zigzag-stitch along edges of front and back catching ends of loop and stuff with fiberfill as you sew.

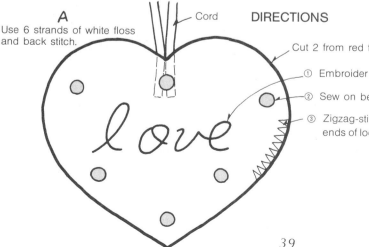

love

Continued on page 75.

Small Pictures ■ Instructions on page 42.

Doll ■ Instructions on page 74.

Small Pictures, *shown on page 40.*

MATERIALS:
White linen for background, 29cm by 35cm each. For A:
Cotton fabric: Checks, 16cm by 19cm; floral print, 8cm by
4.5cm. Sheeting: Gray, 11cm by 12.5cm; white, 11cm by
5cm. For B: Cotton fabric: Checks, 16cm by 19cm; floral

Add 0.5cm for seam allowance all around.
Use 3 strands of floss unless otherwise indicated.

Actual-size Pattern and DIRECTIONS

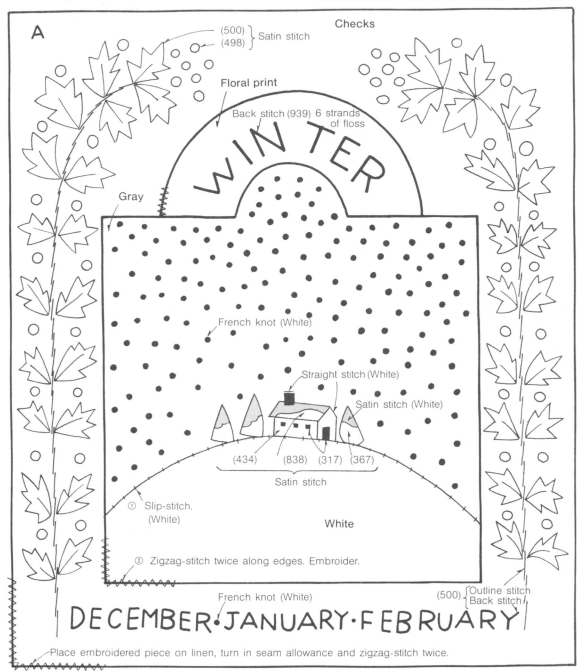

print, 8cm by 4.5cm. Sheeting: Sky blue, 11cm by 13cm; olive green, 11cm by 5cm; green, 11cm by 5cm; white, 6cm by 5cm; blue, 4cm square. DMC six strand embroidery floss, No. 25: For A: One skein of ivy green (500); half skein of scarlet (498); small amount each of white, indigo (939), pistachio green (367), umber (434), beige brown (838), ash gray (317). For B: One skein of emerald green (912); half skein of soft pink (776); small amount each of indigo (823, 311, 322), scarab green (3348), pistachio green (319, 367, 368), emerald green (911), umber (435), smoke gray (644), beige brown (838), lemon yellow (445), saffron (727) and white. Frame, 23.5cm by 29.5cm (inside measurements).

FINISHED SIZE: Same size as frame.

B

Checks

Floral print

Bullion stitch
Satin stitch } (776)

Satin stitch
Outline stitch } (912)

SUMMER

Back stitch
(823)
Use 6
strands of floss.

Sky blue

Satin stitch
Straight
stitch } (322)

Slip-stitch.
(White)

Satin stitch
(644) (910)

Slip-stitch
(367) (311) (3348)

Back stitch
Satin stitch } (838)

Satin stitch
(434)

Satin stitch
(367)

Blue

Outline stitch
Straight
stitch } (434)

Green

JUNE • JULY • AUGUST

Straight stitch (368)

French knot
(727)

Back stitch (319)

(776) (445)
French knot

Olive green

43

Stuffed Tea Set ■ Instructions on page 46.

GIFT

Stuffed Tea Set, *shown on page 44.*

Tea Pot, Sugar Pot and Creamer

MATERIALS:
Floral print, 90cm by 30cm. White sheeting, 50cm by 20cm. White felt, 45cm by 50cm. Polyester fiberfill, 170g.
FINISHED SIZE: See diagrams.

CUTTING DIAGRAMS FOR TEA POT

Add 0.5cm for seam allowance.

Side
Outer piece Print ↕ }
Backing Felt ↔ } Cut 1 each.

14

47

Base
Outer piece Print } Cut 1
Backing Felt } each.

15

Lid
Cut 1 from print.

10

Knob
Cut 1 from white.

3.5

Mouth Cut 2 from print.

0.7 · 3.5 · 0.8 · 1.2
9 · 4 · 0.3
6 · 1.7
4
1.5 · 8

Top of mouth
Cut 1 from white.

1 · 2.5

Border for lid Cut 1 from white.

1.5

28

Handle
Cut 1 from print.

7

15

DIRECTIONS

⑤ Stuff lid with fiberfill,
turn in seam allowance and sew onto body.

④ Stuff with fiberfill.
Run a gathering stitch around top edge and pull thread.

⑦ Run a gathering stitch and pull thread. Sew in place.

⑥ Sew border along seam, Slip-stitch stuffing as you sew.
Lid

2cm

2.5cm

15cm

Side

① Place print on felt and stitch along edges. Machine-stitch vertically at 3.5cm intervals.

② Sew side seam with right sides facing.

3.5cm

16 cm

⑨ Stuff with fiberfill and sew on top.

⑪ Sew pieces together with right sides facing.

⑫ Stuff with fiberfill and sew handle onto body.

⑩ Sew mouth onto body.

⑧ Sew pieces together w right sides facing.

1.5cm

3cm

③ Place print on felt and stitch along edge. Sew base onto body with right sides faci

15cm

46

CUTTING DIAGRAMS FOR SUGAR POT

Add 0.5cm for seam allowance all around.

Side
Outer piece Print ↕
Backing Felt ↔
} Cut 1 each.

7

— 27 —

Lid Cut 1 from print. ↕

10

Base Cut 1 each.

6

Outer piece Print
Backing Felt

Knob
2.5 ↕
Cut 1 from white.

1.5
Border Cut 1 from white. ↕

— 28 —

Handle Cut 2 from white.
2 ↕
6.5

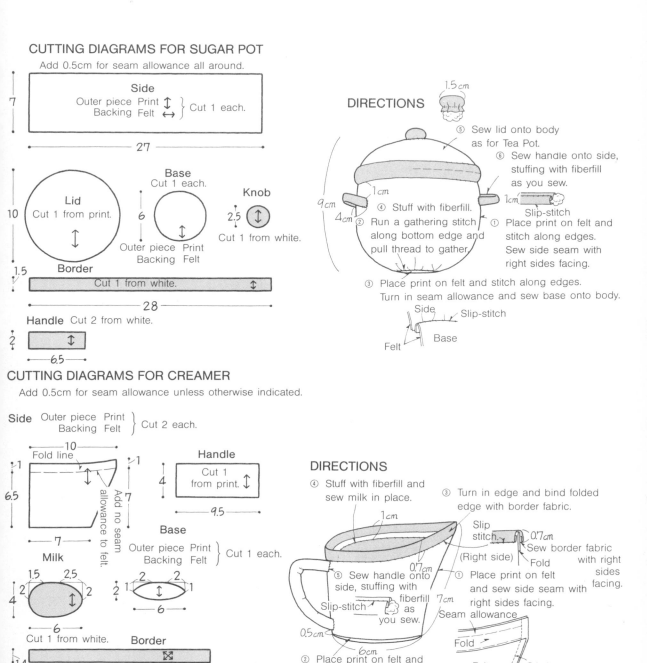

DIRECTIONS

1.5cm

⑤ Sew lid onto body as for Tea Pot.

⑥ Sew handle onto side, stuffing with fiberfill as you sew.

9cm

1cm

④ Stuff with fiberfill.

4cm ② Run a gathering stitch along bottom edge and pull thread to gather.

1cm
Slip-stitch

① Place print on felt and stitch along edges. Sew side seam with right sides facing.

③ Place print on felt and stitch along edges. Turn in seam allowance and sew base onto body.

Side / Slip-stitch
Felt   Base

CUTTING DIAGRAMS FOR CREAMER

Add 0.5cm for seam allowance unless otherwise indicated.

Side Outer piece Print
Backing Felt
} Cut 2 each.

— 10 —
Fold line
1   1
6.5
Add no seam allowance to felt.
7
— 7 —

Handle
4
Cut 1 from print. ↕
— 9.5 —

Base
Outer piece Print
Backing Felt
} Cut 1 each.

Milk
1.5 2.5
2   2
4 ↕
— 6 —
Cut 1 from white.

2 2
2 1 ↕ 1
— 6 —

Border
1.4
— 22 —
Cut 1 from white.

DIRECTIONS

④ Stuff with fiberfill and sew milk in place.

③ Turn in edge and bind folded edge with border fabric.

1cm

Slip stitch

0.7cm
Sew border fabric with right sides facing.

(Right side) Fold

⑤ Sew handle onto side, stuffing with fiberfill as you sew.

① Place print on felt and sew side seam with right sides facing.

Slip-stitch

7cm
Seam allowance

0.5cm

② Place print on felt and sew base onto body with right sides facing.

6cm

0.7cm

Fold
Felt   Stitch

Continued on page 78.

47

MY ROOM

Appliquéd Pillows,
shown on pages 48 & 49.

MATERIALS FOR ONE:
Cotton fabric: For A: Plaid, 71cm by 46cm; floral print, 42cm by 29cm; navy, 29cm square. Fabric for appliqué: Dots, 18cm square; beige, 12cm square; white, 10cm square; dark brown, 10cm by 5cm; sky blue, 6cm square. For B: Plaid, 81cm by 58cm; floral print, 42cm by 11cm;

CUTTING DIAGRAMS AND DIRECTIONS

Add 1cm for seam allowance unless otherwise indicated in parentheses.

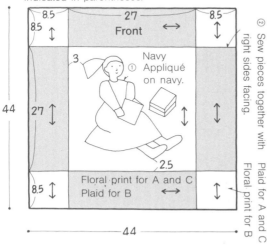

② Sew pieces together with right sides facing.

Plaid for A and C
Floral print for B

Front

Navy
① Appliqué on navy.

Floral print for A and C
Plaid for B

8.5 27 8.5
8.5
3
44
27
2.5
8.5
44

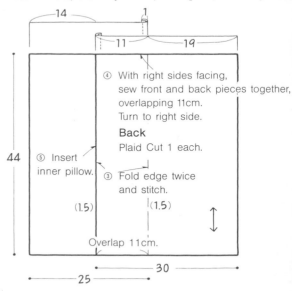

14 1
11 19

④ With right sides facing, sew front and back pieces together, overlapping 11cm. Turn to right side.

Back
Plaid Cut 1 each.

⑤ Insert inner pillow.

③ Fold edge twice and stitch.

(1.5) (1.5)

Overlap 11cm.

44
25 30

Appliqué Pattern

Add no seam allowance to appliqué pieces, but add some allowance to pieces to be overlapped. Glue pieces onto background, then zigzag-stitch with white sewing thread.

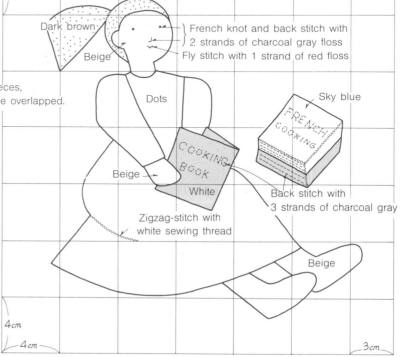

For A:
3cm

Dark brown
Beige

French knot and back stitch with 2 strands of charcoal gray floss
Fly stitch with 1 strand of red floss

Dots

Sky blue
FRENCH COOKING

Beige

COOKING BOOK
White

Back stitch with 3 strands of charcoal gray

Zigzag-stitch with white sewing thread

Beige

4cm
4cm 3cm

navy, 29cm square. Fabric for appliqué: Checks, 22cm by 13cm; white, 15cm square; dots, 15cm by 10cm; sky blue, 10cm square; brown, 11cm by 3cm; dark brown, 10cm by 5cm; beige, 8cm square. For C: Plaid, 71cm by 46cm; floral print, 42cm by 29cm; navy, 29cm square. Fabric for appliqué: Checks, 17cm by 15cm; sky blue, 14cm by 18cm; dark brown, 17cm by 13cm; dots, 12cm square; white, 10cm square; beige, 10cm by 6cm; brown, 8cm square; coffee brown, 6cm by 3cm. Six strand embroidery floss, No. 25: Small amount each of charcoal gray and red for A, charcoal gray, red and sky blue for B, charcoal gray, sky blue, ocher and red for C. Inner pillow, 45cm square.

FINISHED SIZE: 44cm square.

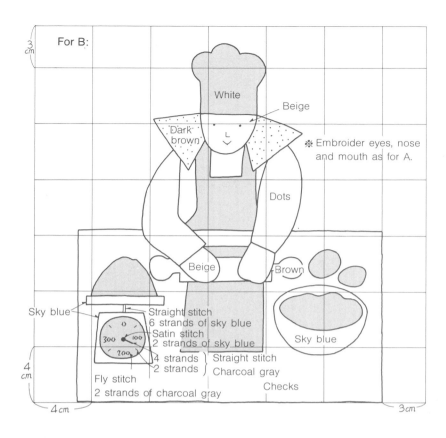

For B:

White

Beige

Dark brown

※ Embroider eyes, nose and mouth as for A.

Dots

Beige

Brown

Sky blue

Straight stitch
6 strands of sky blue
Satin stitch
2 strands of sky blue

Sky blue

4 strands   Straight stitch
2 strands   Charcoal gray

Fly stitch
2 strands of charcoal gray

Checks

4cm

For C:

Embroider eyes, nose
※ and mouth as for A.

Beige

Brown

French knot
2 strands of sky blue

Beige

Dots

White

Sky blue

Beige

Coffee brown

Back stitch
3 strands of ocher

Zigzag-stitch.
White

Checks

Dark brown

51

Practical Bags

Travel Bags and Cases,
shown on pages 52 & 53.

Travel Bag

MATERIALS:

Blue cotton bonded fabric, 80cm by 78cm. 40cm long zipper. 4 emblems.

FINISHED SIZE: See diagram.

CUTTING DIAGRAMS

Add seam allowance indicated in

(0.7)

4.5

4.5 Place for emblem

5.5

9.5 4.3

(0.5)

35

(0.5)

Front and Back
Cut 1.

Fold

42

Add 0.5cm seam allowance around edge.

7

4

25

Gusset
Cut 2.

1

2 6

6

17.5

Stay
Cut 4.
Add 1cm seam allowance around edge.

4   3

4.5 No seam allowance

Tab
Cut 2.
(0.5)

4

(0.5)
6

No seam allowance

※ Peel off sponge from wrong side of fabric for stay and tab.

Handle Cut 2.
Add 1cm seam allowance around edge.

Seam allowance

6

47

DIRECTIONS

② Sew on zipper.

Zipper

1cm

1.5 0.5
cm   cm

Zigzag stitch.

① Sew on emblems.

④ Sew handles.

3cm

Turn in seam allowance 2cm and stitch.

10cm

Tab

10 cm

2.3 cm

⑤ Sew handles with stay on top.

③
With right sides facing and folded tab in between sew gusset onto Zigzag-stitch.

42cm

17.5 cm

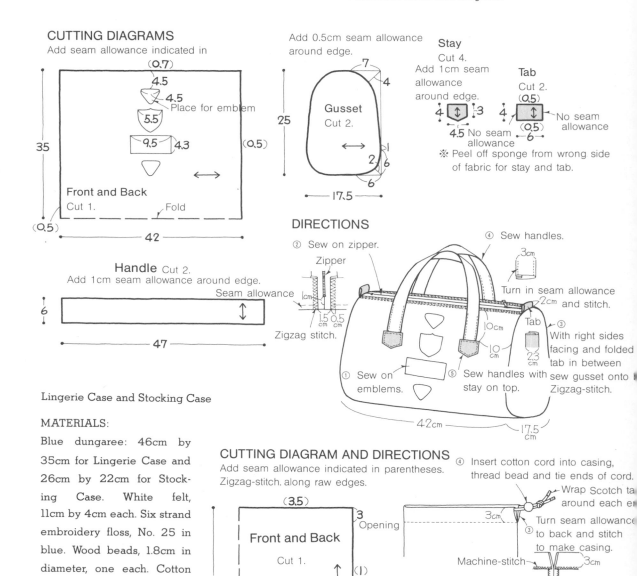

Lingerie Case and Stocking Case

MATERIALS:

Blue dungaree: 46cm by 35cm for Lingerie Case and 26cm by 22cm for Stocking Case. White felt, 11cm by 4cm each. Six strand embroidery floss, No. 25 in blue. Wood beads, 1.8cm in diameter, one each. Cotton cord, 0.5cm in diameter, 64cm long for Lingerie Case and 44cm for Stockings Case.

FINISHED SIZE: Lingerie Case, 22cm by 30cm. Stocking Case, 12cm by 18cm.

CUTTING DIAGRAM AND DIRECTIONS

Add seam allowance indicated in parentheses. Zigzag-stitch. along raw edges.

(3.5)

3
Opening

Front and Back

Cut 1.

(1)

30

Fold

① Embroider and appliqué on front.

5.5

(1)

1.5

4

22

② Fold in half with right sides facing and sew side and bottom seams.

④ Insert cotton cord into casing, thread bead and tie ends of cord.

3cm

Wrap Scotch tape around each end

Turn seam allowance ③ to back and stitch to make casing.

Machine-stitch

3cm

CUTTING DIAGRAM FOR STOCKING CASE

Add seam allowance indicated in parentheses. Zigzag-stitch along raw edges.

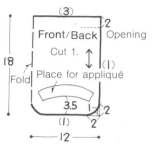

Make as for Lingerie Case.

Actual-size Embroidery and Appliqué Patterns

Use 6 strands of floss.

Back stitch

Felt

LINGERIE

STOCKINGS

CLOTHES

Shirt Case

MATERIALS:

Blue dungaree and red checks, 76cm by 37cm each. White felt, 10cm by 3cm. Six strand embroidery floss, No. 25 in blue. Snap fasteners, 1cm in diameter, 2 pairs.

FINISHED SIZE: See diagram.

CUTTING DIAGRAM

Add seam allowance indicated in parentheses.

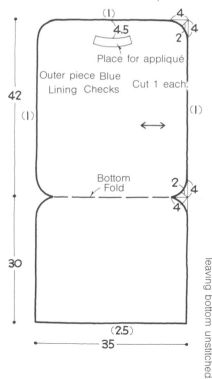

Outer piece Blue
Lining Checks   Cut 1 each.

Place for appliqué

Bottom Fold

DIRECTIONS

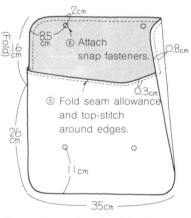

⑥ Attach snap fasteners.

⑤ Fold seam allowance and top-stitch around edges.

① Embroider and appliqué in place.

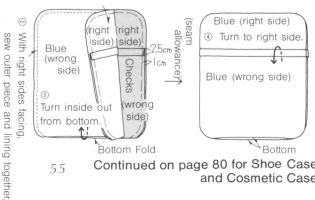

② With right sides facing, sew outer piece and lining together, leaving bottom unstitched.

③ Turn inside out from bottom.

Blue (wrong side)

Checks (wrong side)

Bottom Fold

④ Turn to right side.

Blue (right side)

Blue (wrong side)

Bottom

55

Continued on page 80 for Shoe Case and Cosmetic Case.

Day Pack and Waist Pouch,

shown on page 56.

Waist Pouch

MATERIALS:

Blue quilted fabric, 90cm by 17cm. Cotton fabric: Sky blue, 12cm square; plaid, 14cm by 7cm; navy, 8cm by 6cm; red, 7cm square; purple, 6cm by 4cm; beige, 5cm square; light brown, 4cm square; white, 3cm square. 3 black beads. Navy zipper, 27cm long. One pair of navy buckles. Unbleached cotton tape, 2.5cm by 87cm. Glue.

FINISHED SIZE: See diagram.

Actual-size Appliqué Patterns for Lid and Front

Add no seam allowance to appliqué pieces, but add extra allowance to pieces to be overlapped.

Glue and zigzag-stitch along edges with white sewing thread.

Lid

Sky blue

Front

Light brown

Zigzag-stitch. White

Beige

Beads

Light brown

Red

White

Plaid

Sky blue

Purple

Plaid

Navy

Machine-stitch. White

58

CUTTING DIAGRAMS

Add 1cm for seam allowance all around.

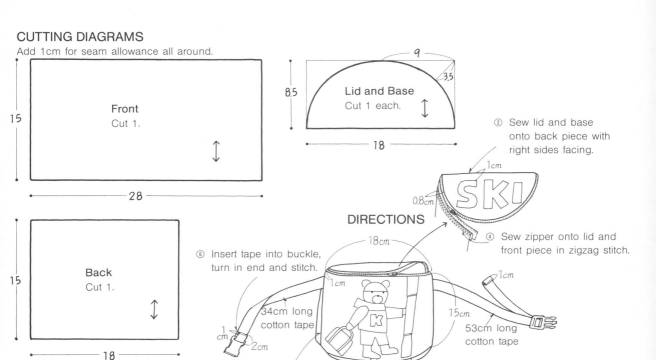

Front
Cut 1.

15

28

Lid and Base
Cut 1 each.

9

3.5

8.5

18

② Sew lid and base onto back piece with right sides facing.

Back
Cut 1.

15

18

1cm

0.8cm

SKI

DIRECTIONS

④ Sew zipper onto lid and front piece in zigzag stitch.

⑥ Insert tape into buckle, turn in end and stitch.

18cm

1cm

1cm

K

34cm long cotton tape

1cm

2cm

15cm

53cm long cotton tape

③ Sew side seams with right sides facing and 1cm of cotton tape in between.

0.5cm

⑤ Sew front and base together.

① Appliqué on front and lid.

Day Pack

MATERIALS:

Blue quilted fabric, 43cm by 80cm. Cotton fabric: Beige, 12cm by 4cm; light brown, 6cm square; plaid, 10cm by 5cm; white, 9cm by 7cm; red, 6cm square; purple, 6cm by 3cm; navy, 4cm square. Six strand embroidery floss, No. 25: One skein of navy and small amount of brown. 2 black beads. Navy snap fasteners, 1.2cm in diameter, 2 pairs. Unbleached cotton tape, 2.5cm by 109cm. Glue.

FINISHED SIZE: See diagram.

CUTTING DIAGRAM

Add 1cm for seam allowance unless otherwise indicated in parentheses.

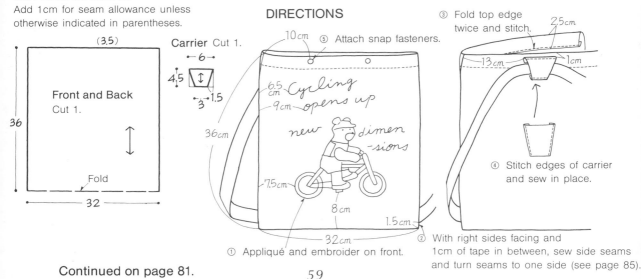

(3.5)

Front and Back
Cut 1.

36

32

Fold

Carrier Cut 1.

6

4.5

3

1.5

DIRECTIONS

⑤ Attach snap fasteners.

10cm

③ Fold top edge twice and stitch.

2.5cm

13cm

1cm

6.5 cm Cycling

9cm opens up

new dimen -sions

36cm

7.5cm

8cm

1.5cm

④ Stitch edges of carrier and sew in place.

① Appliqué and embroider on front.

32cm

② With right sides facing and 1cm of tape in between, sew side seams and turn seams to one side (see page 85).

Continued on page 81.

Linen Bag ■ Instructions on page 62.

Patchwork Bag ■ Instructions on page 62.

Linen Bag, *shown on page 60.*

MATERIALS:

Light brown linen and brown floral print, 74cm by 35cm each. Dark brown sheeting, 52cm by 9cm. Six strand embroidery floss, No. 25 in dark brown, green and unbleached. 4 buttons, 1.1cm in diameter.

FINISHED SIZE: See diagram.

CUTTING DIAGRAMS

Add 1cm for seam allowance unless otherwise indicated in parentheses.

Front and Back

Outer piece Linen
Lining Print

Cut 2 each.

(2)

32

13

5.5

12

35

Handle

Cut 2 from sheeting.

(3)

46 — Fold line

2.5

(3)

DIRECTIONS

③ Make handles.

④ Insert lining into outer bag with wrong sides facing and stitch along top edges catching ends of handles.

3cm

10cm

⑤ Sew on buttons.

① Embroider on front.

32 cm

6.5 cm

② With right sides facing, sew front and back together. Sew pieces of lining in same manner.

35cm

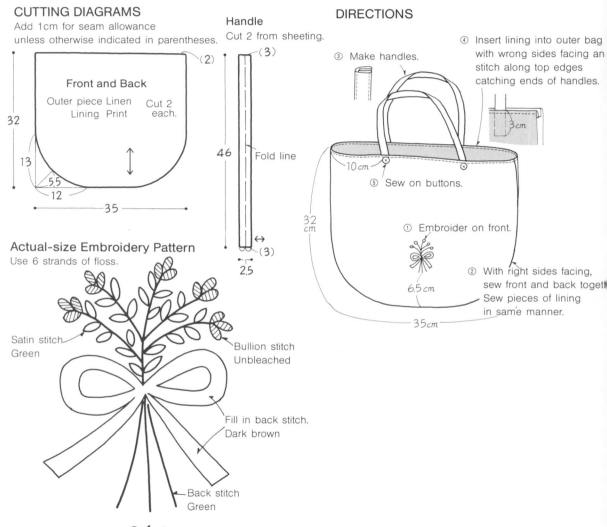

Actual-size Embroidery Pattern

Use 6 strands of floss.

Satin stitch
Green

Bullion stitch
Unbleached

Fill in back stitch.
Dark brown

Back stitch
Green

Patchwork Bag, *shown on page 61.*

MATERIALS:

Black cotton broadcloth, 79cm by 55cm. Cotton fabric: Black print (B), 49cm by 8.5cm; black print (A) and reddish brown with dots, 49cm by 6.5 cm each; black with dots, navy with dots, black and white checks, beige and black checks, 49cm by 4.5 cm each. Thick quilt batting, 83cm by 30cm. Elastic tape, 0.6cm by 17cm.

FINISHED SIZE: See diagram.

CUTTING DIAGRAMS

Add seam allowance indicated in parentheses.

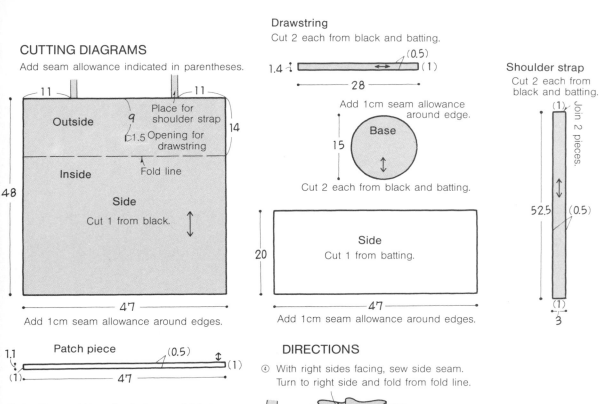

Drawstring
Cut 2 each from black and batting.

(0.5)
1.4
28
(1)

Shoulder strap
Cut 2 each from black and batting.

(1) Join 2 pieces.
52.5 (0.5)
(1)
3

11 — 11
Outside
9 Place for shoulder strap
1.5 Opening for drawstring
14
48
Inside
Fold line
Side
Cut 1 from black.
47
Add 1cm seam allowance around edges.

Add 1cm seam allowance around edge.
Base
15

Cut 2 each from black and batting.

Side
Cut 1 from batting.
20
47
Add 1cm seam allowance around edges.

Patch piece
1.1
(0.5)
(1)
(1)
47

Color Key and Required pieces of fabric

Color		Amount
	Black print (A)	Cut 3.
	Black with dots	" 2.
	Black print (B)	" 4.
	Reddish brown with dots	" 3.
	Black and white checks	" 2.
	Navy with dots	" 2.
	Beige and black checks	" 2.

DIRECTIONS

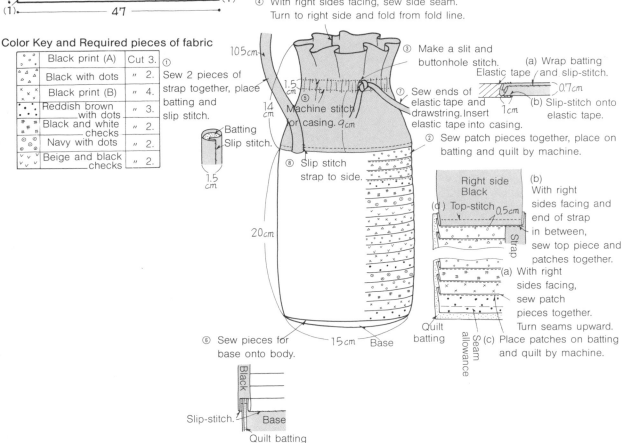

④ With right sides facing, sew side seam. Turn to right side and fold from fold line.

③ Make a slit and buttonhole stitch.

(a) Wrap batting and slip-stitch.
Elastic tape
0.7cm
(b) Slip-stitch onto elastic tape.
1cm

⑦ Sew ends of elastic tape and drawstring. Insert elastic tape into casing.

① Sew 2 pieces of strap together, place batting and slip stitch.

105cm
1.5cm
14cm
⑤ Machine stitch for casing. 9cm
20cm
15cm Base

Batting
Slip stitch.
1.5cm

⑧ Slip stitch strap to side.

② Sew patch pieces together, place on batting and quilt by machine.

Right side Black
(d) Top-stitch 0.5cm
Strap

(b) With right sides facing and end of strap in between, sew top piece and patches together.

(a) With right sides facing, sew patch pieces together. Turn seams upward.

(c) Place patches on batting and quilt by machine.

Quilt batting
Seam allowance

⑥ Sew pieces for base onto body.

Black
Slip-stitch.
Base
Quilt batting

Quilted Shoulder Bag ■ Instructions on page 65.

64

Quilted Shoulder Bag,
shown on page 64.

MATERIALS:

Nylon bonded fabric: Green, 90cm by 30cm; navy, 90cm by 20cm. Polyurethane foam, 90cm by 20cm. Zippers, 30cm and 24cm long each. Emblem.

FINISHED SIZE: See diagram.

CUTTING DIAGRAMS

Add 0.5cm for seam allowance unless otherwise indicated in parentheses.

Front A
Outer piece Green
Lining Navy } Cut 1 each.
Interlining Polyurethane

11
24

Front B
Cut 1 each as for Front A.
2
24

Top and Back
Cut 2 each as for Front A.
14
18

Top and Bottom
Cut 2 each as for Front A.
5
18
5 2

Gusset A Cut 2 each as for Front A.
1.5
30

Gusset B
Cut 1 each as for Front A.
4
34
Add 1cm seam allowance around edges.

Strap Cut 1 from green.
Join.
3
56.5   56.5

DIRECTIONS

⑨ Place green on polyurethane.
Sew front and gussets together with right sides facing.
Sew navy for lining onto wrong side.

⑩ Sew back and gussets as for step ⑨.

⑥ Turn in seam allowance and stitch.

⑦ Place green on polyurethane. Sew gusset and top together with right sides facing and end of strap in between.

⑤ Sew zipper as for step ②.

③ Place top and back pieces on polyurethane. Sew back and front B together with right sides facing. Sew navy for lining onto wrong side.

② Sew zipper in place.
(b) Turn in seam allowance and machine-stitch.
(a) With right sides facing, sew navy and zipper together.

Front B
Green Front A (Right side)
Navy (Wrong side)

Seam allowance
Polyurethane
Stitch.
Green

① Place green Front A and B on polyurethane and machine-stitch at 1cm intervals.

⑪ Attach emblem.

④ Sew base onto front A as for step ③.

Gusset A (Right side) Strap
Machine-stitch.
Green (Wrong side)
Gusset B
Polyurethane

⑧ Turn in seam allowance of navy and sew onto wrong side. Slip-stitch.
(b) Slip-stitch.
(a) Machine-stitch.
Gusset B Navy
Gusset A

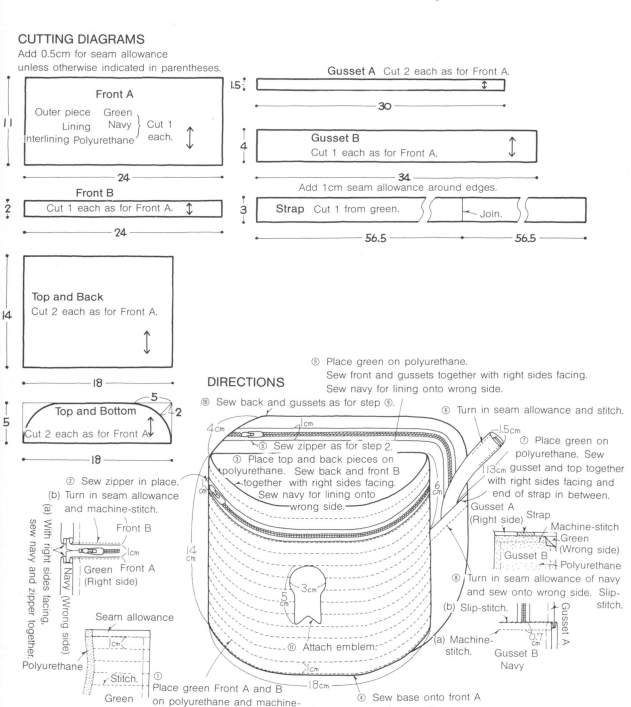

65

Wall Hanging, *shown on Back Cover.*

Appliqué Pattern

Add 0.5cm for seam allowance to appliqué pieces unless otherwise indicated.
Use 2 strands of floss in matching color and slip stitch.
Embroider with 3 strands of floss (841) and back stitch unless otherwise indicated.

Add 2cm all around to background fabric and 1cm each for joining.

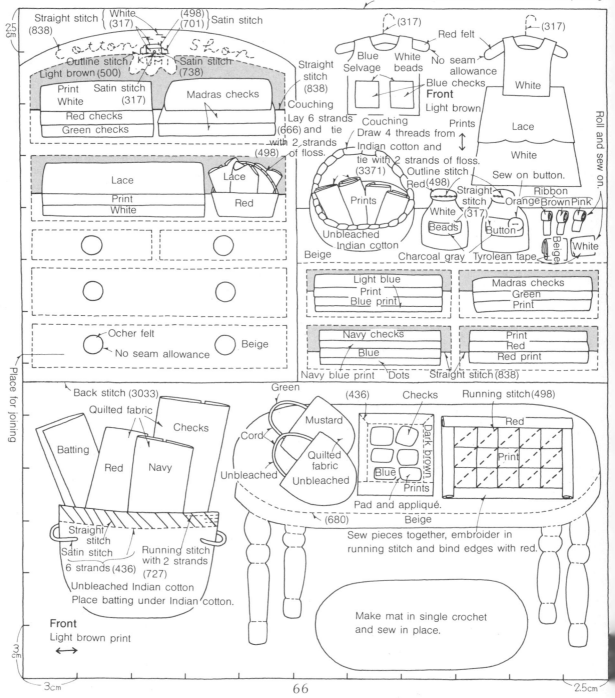

Straight stitch (838)
White (317) (498) (701) Satin stitch
Outline stitch KUMI Satin stitch (738)
Light brown (500)
Print White Satin stitch (317)
Madras checks
Red checks
Green checks
Straight stitch (838)
Couching
Lay 6 strands (666) and tie with 2 strands (498) of floss.

Lace
Lace
Print White
Red

Ocher felt
No seam allowance
Beige

(317)
Blue Selvage White beads
Red felt
No seam allowance
Front
Light brown
Prints
Blue checks

Couching
Draw 4 threads from Indian cotton and tie with 2 strands of floss. (3371)
Outline stitch Red (498)
Straight stitch (317)
Prints
White Beads
Button
Unbleached Indian cotton
Beige
Charcoal gray
Tyrolean tape

(317)
White
Lace
White
Sew on button.
Ribbon
Orange Brown Pink
Beige
White

Light blue
Print
Blue print
Madras checks
Green
Print

Navy checks
Blue
Print
Red
Red print

Navy blue print
Dots
Straight stitch (838)

Back stitch (3033)
Quilted fabric
Checks
Batting
Red
Navy
Straight stitch
Satin stitch
6 strands (436)
Running stitch with 2 strands (727)
Unbleached Indian cotton
Place batting under Indian cotton.

Green
Cord
Mustard
Quilted fabric
Unbleached
Unbleached
(680)

(436)
Checks
Dark brown
Blue
Prints
Pad and appliqué.
Beige

Running stitch (498)
Red
Print

Sew pieces together, embroider in running stitch and bind edges with red.

Make mat in single crochet and sew in place.

Front
Light brown print

25cm
3cm
3cm
2.5cm

Place for joining
Roll and sew on.

MATERIALS:

Cotton fabric: Red, 50cm square; light brown print, 58cm by 37cm; beige, 58cm by 22cm; light brown, 25cm by 10cm; scraps of white, blue, green, dark brown, light blue, dots, charcoal gray; scraps of red, green, white, blue, pink, yellow, navy prints; scraps of red, green, brown plaids, Madras checks, blue and white, navy and gray gingham checks. Scrap of lace. Quilted fabric: Scraps of red and green checks, red, navy, mustard and unbleached. Scrap of unbleached Indian cotton fabric. DMC six strand embroidery floss, No. 25: One skein of beige brown (841); half skein each of umber (436) and old gold (680); small amount each of scarlet (498), ivy green (500), brilliant green (701), ash gray (317), saffron (727), poppy (666), dark brown (3033), umber (738), sepia (3371), beige brown (838) and white. Tyrolean tape in beige and white, 1.2cm by 7cm each. Ribbon in orange, brown and pink, 0.5cm by 6cm each. One button, 0.9cm in diameter. Two white beads (small). Green and unbleached cords, 6cm long each. Scraps of ocher and red felt. Scrap of lace. Thin quilt batting, 45cm square. Embroidery thread, No. 4: One skein of brown and small amount of unbleached. Steel crochet hook size 3mm.

FINISHED SIZE: 36.5cm by 42.5cm.

CUTTING DIAGRAM AND DIRECTIONS
Add 1cm for seam allowance.

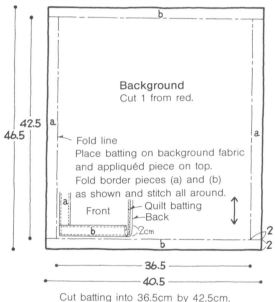

Background
Cut 1 from red.

Fold line
Place batting on background fabric and appliquéd piece on top.
Fold border pieces (a) and (b) as shown and stitch all around.

Front
— Quilt batting
— Back
2cm

42.5
46.5

36.5
40.5

Cut batting into 36.5cm by 42.5cm.

Mat
Use cotton embroidery thread, No. 4 in brown and 3mm steel crochet hook.

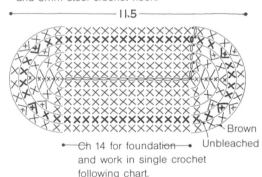

11.5

Brown
Unbleached

Ch 14 for foundation and work in single crochet following chart.

Patchwork Bag, *shown on page 17.*

MATERIALS:

Dark brown sheeting, 90cm by 70cm. Madras checks, 70cm by 7cm. Brown print, 34cm by 16cm. Thick quilt batting, 68cm by 46cm. Six strand embroidery floss, No. 25: 2 skeins of dark brown. One pair of snap fasteners, 1.5cm in diameter.

FINISHED SIZE: 33cm wide and 30cm deep.

CUTTING DIAGRAMS

Add 0.5cm for seam allowance all around.

Patch piece

Cut 15 each from dark brown and checks.
Cut 2 from dark brown and 3 from checks.

5 — 6

5 — 3

4.5 — Cut 2 from print. — 33

3.5 — Cut 1 from print. — 33

2.5 — Cut 3 from dark brown. — 33

Stay
Cut 2 from dark brown.

4 — 4.5

27.5

Back
Cut 1 from dark brown.
Lining
Cut 2 each from dark brown.
Quilt batting
Cut 2.

9.5

5

8

33

Border strip
Cut 1 each from dark brown and batting.

Add no seam allowance to batting.

7 — 66

No seam allowance

Handle Cut 2 each from dark brown and batting.

5 — 55

Add no seam allowance to batting.

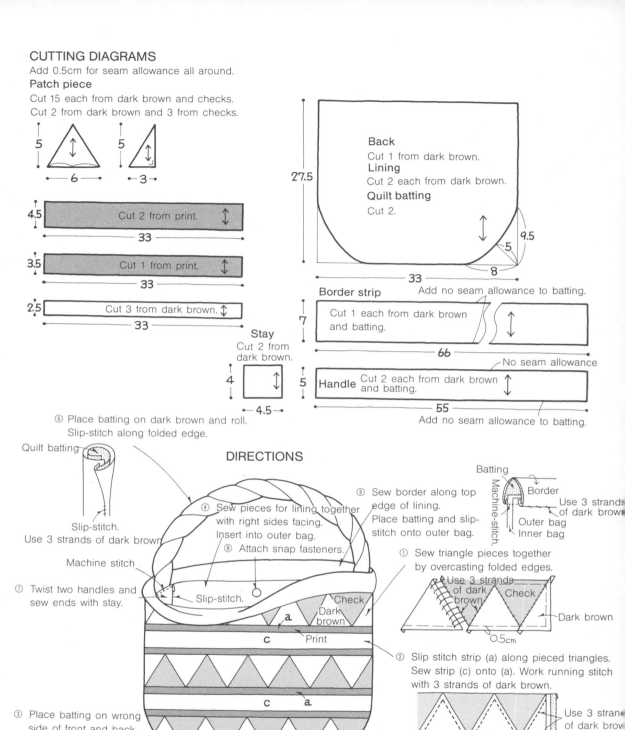

⑥ Place batting on dark brown and roll.
Slip-stitch along folded edge.

Quilt batting

Slip-stitch.
Use 3 strands of dark brown

DIRECTIONS

Machine stitch

⑦ Twist two handles and sew ends with stay.

Slip-stitch.

④ Sew pieces for lining together with right sides facing.
Insert into outer bag.
⑧ Attach snap fasteners.

⑤ Sew border along top edge of lining.
Place batting and slip-stitch onto outer bag.

Batting
Border
Use 3 strands of dark brown
Outer bag
Inner bag
Machine-stitch.

① Sew triangle pieces together by overcasting folded edges.

Use 3 strands of dark brown
Check
Dark brown

0.5cm

② Slip stitch strip (a) along pieced triangles.
Sew strip (c) onto (a). Work running stitch with 3 strands of dark brown.

Use 3 strands of dark brown

c
a
Quilt batting

Check
Dark brown
Print
a
c
c   a
b
c

③ Place batting on wrong side of front and back.
With right sides facing, sew front and back together.
Turn to right side.

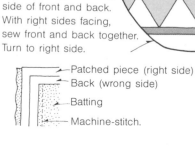

Patched piece (right side)
Back (wrong side)
Batting
Machine-stitch.

Continued from page 11.

Hot Mitt

MATERIALS FOR ONE:

Sheeting, 50cm by 52cm, red for A, yellow for B and blue for C. White sheeting, 13cm by 6cm. Thick quilt batting, 46cm by 26cm. Six strand embroidery floss, No. 25: Small amount each of moss green, red and charcoal gray for A, lemon yellow, olive green and charcoal gray for B, and blue, blue green and charcoal gray for C.

FINISHED SIZE: 25cm deep.

CUTTING DIAGRAMS

Add 1cm for seam allowance unless otherwise indicated.

Strip for binding

Pattern

Add 1cm for seam allowance unless otherwise indicated.

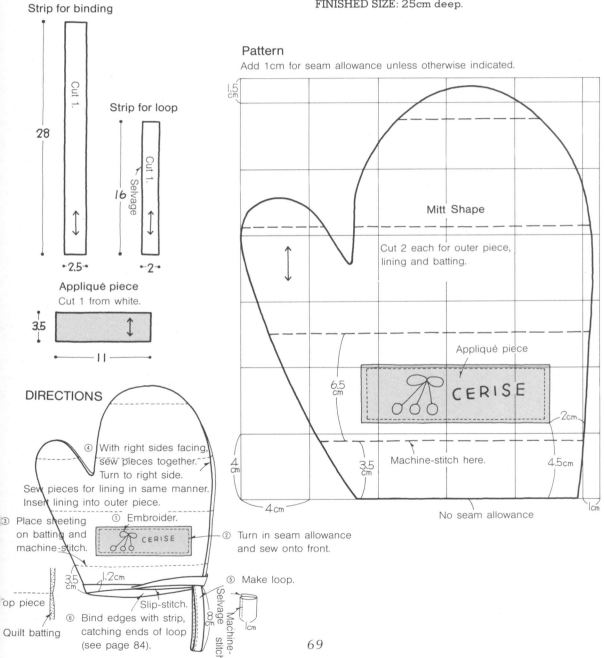

Cut 1.

28

2.5

Strip for loop

Cut 1.

Selvage

16

2

Appliqué piece
Cut 1 from white.

3.5

11

Mitt Shape

Cut 2 each for outer piece, lining and batting.

Appliqué piece

CERISE

1.5 cm

6.5 cm

4 cm

3.5 cm

4 cm

Machine-stitch here.

2cm

4.5cm

1cm

No seam allowance

DIRECTIONS

④ With right sides facing, sew pieces together. Turn to right side.
Sew pieces for lining in same manner. Insert lining into outer piece.

③ Place sheeting on batting and machine-stitch.

① Embroider.

CERISE

② Turn in seam allowance and sew onto front.

⑤ Make loop.

3.5 cm

1.2cm

Slip-stitch.

Selvage 8cm

Machine-stitch.

1cm

Top piece

Quilt batting

⑥ Bind edges with strip, catching ends of loop (see page 84).

69

Night Dress and Slippers,
shown on page 29.

Night Dress

MATERIALS:

Unbleached sheeting, 90cm by 3m. Six strand embroidery floss, No.25: Small amount of dark green. 3 gray sheel buttons, 1.2cm in diameter.

FINISHED SIZE: Bust, 120cm. Length, 122cm. Width from center back to cuff, 65cm.

CUTTING DIAGRAMS
Add seam allowance indicated in parentheses.

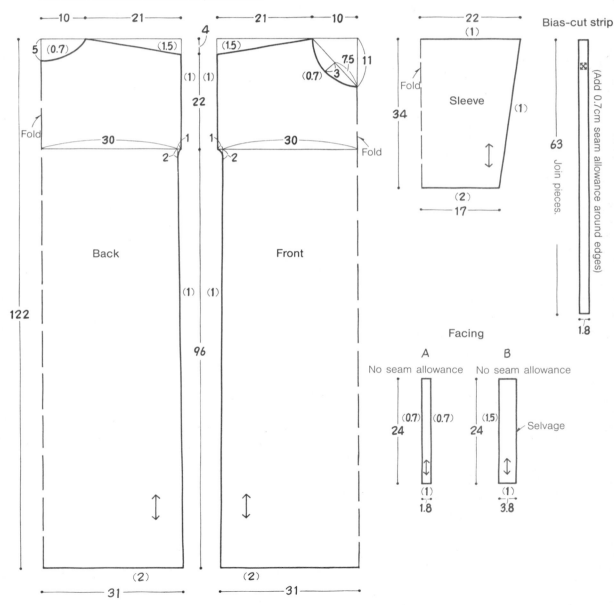

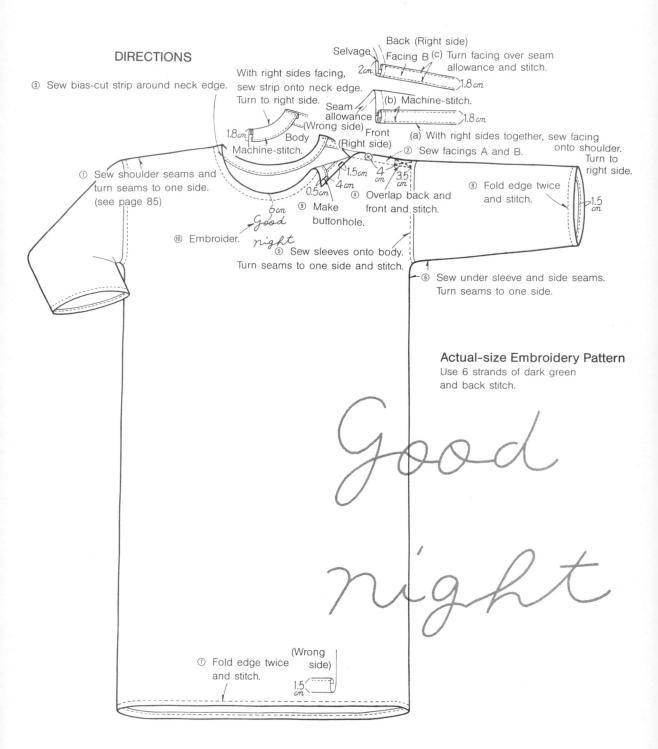

DIRECTIONS

③ Sew bias-cut strip around neck edge.

With right sides facing, sew strip onto neck edge. Turn to right side.

Back (Right side)

Selvage

Facing B (c) Turn facing over seam allowance and stitch.

2cm

1.8cm

(b) Machine-stitch.

Seam allowance (Wrong side)

1.8cm

1.8cm

Body

Machine-stitch.

Front (Right side)

(a) With right sides together, sew facing onto shoulder. Turn to right side.

② Sew facings A and B.

① Sew shoulder seams and turn seams to one side. (see page 85)

1.5cm 4cm 3.5cm

0.5cm 4cm

④ Overlap back and front and stitch.

⑧ Fold edge twice and stitch.

1.5cm

6cm *Good*

⑨ Make buttonhole.

⑩ Embroider. *night*

⑤ Sew sleeves onto body. Turn seams to one side and stitch.

⑥ Sew under sleeve and side seams. Turn seams to one side.

Actual-size Embroidery Pattern
Use 6 strands of dark green and back stitch.

Good

night

(Wrong side)

⑦ Fold edge twice and stitch.

1.5cm

Slippers

MATERIALS:

Sheeting: Unbleached, 84cm by 36cm; dark green, 10cm square. Quilt batting, 84cm by 18cm. Iron-on interfacing, 25cm by 20cm. Four gray shell buttons, 1.2cm in diameter.
FINISHED SIZE: 23cm.

PATTERN AND CUTTING DIAGRAM
Add 1cm for seam allowance all around.

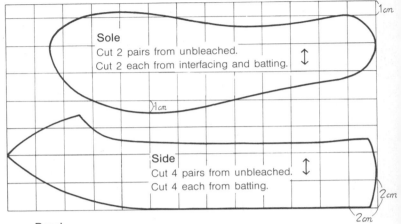

1cm

Sole
Cut 2 pairs from unbleached.
Cut 2 each from interfacing and batting.

1 cm

Side
Cut 4 pairs from unbleached.
Cut 4 each from batting.

2cm

2cm

Band
Cut 2 from green.
Fold.

1.5
7.5

DIRECTIONS

① Place unbleached sheeting on batting.
With right sides facing,
sew pieces for side at center front and back.
Turn to right side.

② Sew pieces for lining in same manner.

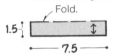

④ Place sole on batting.
Sew pieces for sole and side together with right sides facing.

⑥ With right sides facing, fold strip in half and stitch. Turn to right side and slip stitch opening closed. Sew band in place with button on each end.

Opening for turning

⑤ Press iron-on interfacing onto sole.
Sew sole onto side piece along edge.

③ Sew outer piece and lining together with right sides facing.
Turn to right side.

Side
Slip-stitch.
Sole
Sole
Interfacing
Batting

Gift Box and Cards,
shown on page 37.

Card
MATERIALS FOR ONE:
Off-white cardboard, 18cm by 12.5cm. Glue. For A: Light brown burlap, 12cm square. Unbleached cord, 32cm long. Tag. For B: Polyester fiberfill, 14cm by 1.5cm. Red cord, 50cm long. Six golden beads, 0.3cm in diameter. For C: Off-white Indian cloth, 5cm by 4cm. Six strand embroidery floss, No. 25 in green and pink.

CUTTING DIAGRAM AND DIRECTIONS

For A:

① Fold burlap as shown and tie with cord.

5

12.5

3.5

② Write letters on tag and attach to cord.

③ Glue burlap onto front of card.

FOR YOU

2

3.5

Fold

9

For B:

② Tie fiberfill with cord.

① Make circle with polyester fiberfill.

2 cm

1.5 cm

③ Glue beads.

④ Glue circle onto center of card.

For C:

② Draw one thread each horizontally and vertically. Glue cloth onto card.

① Cross-stitch with 3 strands of pink and green floss.

Green   Pink

4 = 20 squares

Indian cloth

25 squares

Oval Gift Boxes

MATERIALS:

Oval box, 10.5cm by 8cm for A and 13cm by 9.5cm for B. Glue. For A: White flannel, 14cm by 6cm. Golden tape, 0.6cm by 20cm. Golden beads (small), 24 pieces. For B: Light brown burlap, 12cm square. Gold lamé thread, 64cm long. Tag.

Actual-size Pattern

Cut out 2 pieces adding 0.5cm seam allowance.

A

DIRECTIONS

① Sew pieces together with right sides facing.

③ Sew on beads.

④ Glue 6cm-long tape onto front.

2.5 cm

② Make a slit and turn inside out. Stuff with fiberfill and slip-stitch slit closed.

⑤ Tie tape into bow and glue.

⑥ Glue finished heart onto center of lid.

* Make B as for Card A, but tie parcel with 2 strands of gold lamé thread. Glue to center of lid.

Doll, shown on page 41.

MATERIALS:

Red and green plaid, 29cm by 50cm. White sheeting, 32cm by 15cm. White lace, 13cm by 31cm. Cotton embroidery thread, No.4: 1-1/2 skeins of brown. Six strand embroidery floss, No. 25 in charcoal gray, red and brown. Red pencil. White ribbon, 1.8cm by 60cm. Elastic tape, 16cm long.

FINISHED SIZE: 23cm tall.

CUTTING DIAGRAMS

Add 0.5cm for seam allowance unless otherwise indicated in parentheses.

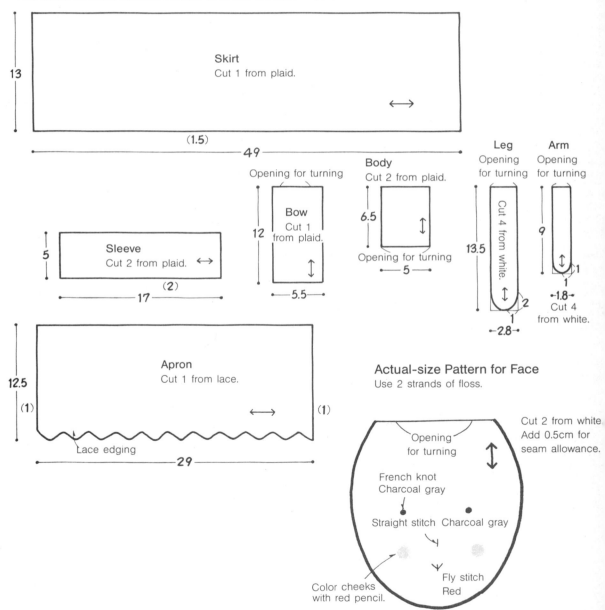

Skirt
Cut 1 from plaid.

13

(1.5)

49

Sleeve
Cut 2 from plaid.

5

17

(2)

Opening for turning

Bow
Cut 1 from plaid.

12

5.5

Body
Cut 2 from plaid.

6.5

Opening for turning

5

Leg
Opening for turning

Cut 4 from white.

13.5

2

1

2.8

Arm
Opening for turning

9

1

1

1.8

Cut 4 from white.

Apron
Cut 1 from lace.

12.5

(1)

(1)

Lace edging

29

Actual-size Pattern for Face

Use 2 strands of floss.

Opening for turning

Cut 2 from white
Add 0.5cm for seam allowance.

French knot
Charcoal gray

Straight stitch Charcoal gray

Fly stitch
Red

Color cheeks
with red pencil.

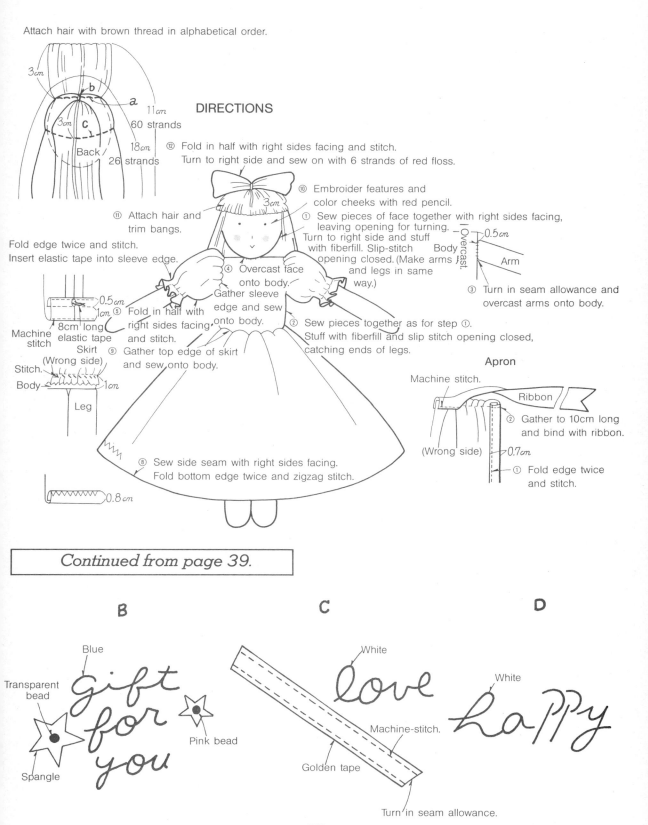

Attach hair with brown thread in alphabetical order.

3cm

b

a

11cm 60 strands

3cm c

Back 18cm 26 strands

DIRECTIONS

⑫ Fold in half with right sides facing and stitch.
Turn to right side and sew on with 6 strands of red floss.

⑩ Embroider features and color cheeks with red pencil.

3cm

⑪ Attach hair and trim bangs.

① Sew pieces of face together with right sides facing, leaving opening for turning.
Turn to right side and stuff with fiberfill. Slip-stitch opening closed. (Make arms and legs in same way.)

0.5cm

Body

Arm

Overcast

Fold edge twice and stitch.
Insert elastic tape into sleeve edge.

0.5cm

1cm ⑤

Machine stitch

8cm long elastic tape

Skirt ⑨

Stitch.

Body

(Wrong side)

Leg

④ Overcast face onto body.
Gather sleeve edge and sew onto body.

⑤ Fold in half with right sides facing and stitch.

⑨ Gather top edge of skirt and sew onto body.

② Sew pieces together as for step ①.
Stuff with fiberfill and slip stitch opening closed, catching ends of legs.

③ Turn in seam allowance and overcast arms onto body.

1cm

Apron

Machine stitch.

Ribbon

② Gather to 10cm long and bind with ribbon.

(Wrong side)

0.7cm

① Fold edge twice and stitch.

⑧ Sew side seam with right sides facing.
Fold bottom edge twice and zigzag stitch.

0.8cm

Continued from page 39.

B

Blue

Transparent bead

gift for you

Pink bead

Spangle

C

White

love

Machine-stitch.

Golden tape

Turn in seam allowance.

D

White

happy

Covered Notebooks and Pencil Case, *shown on page 45.*

Large and Small Notebooks

MATERIALS:

Notebooks: Large, 21cm by 29.5cm; small, 13cm by 18.5cm. For Large Notebook: Black and unbleached checks, 90cm by 34cm. Navy felt, 10cm by 2.5cm. For Small Notebook: Unbleached sheeting, 30cm by 23cm. Black and beige checks, 22cm by 16.5cm. Navy felt, 11cm by 2.5cm. Glue.

FINISHED SIZE: Same size as notebook.

CUTTING DIAGRAM

Add no seam allowance unless otherwise indicated.
Add margin indicated in parentheses.
Sizes and materials indicated in brackets are for Small Notebook.

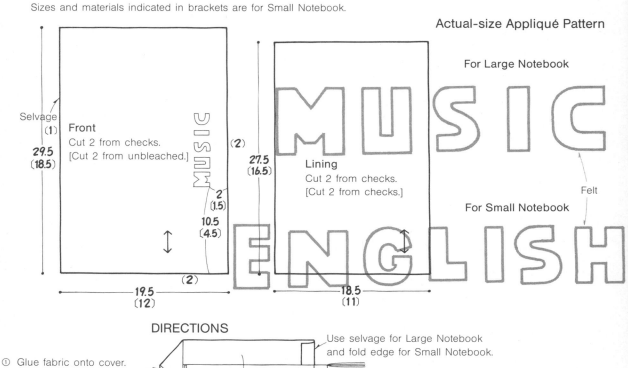

Actual-size Appliqué Pattern

For Large Notebook

For Small Notebook

Felt

Front
Cut 2 from checks.
[Cut 2 from unbleached.]

Selvage
(1)
29.5
(18.5)
2
(1.5)
10.5
(4.5)
(2)
19.5
(12)

Lining
Cut 2 from checks.
[Cut 2 from checks.]

(2)
27.5
(16.5)
18.5
(11)

DIRECTIONS

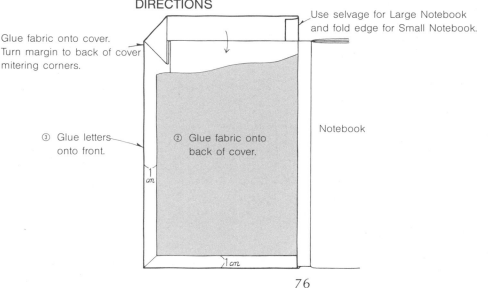

Use selvage for Large Notebook and fold edge for Small Notebook.

① Glue fabric onto cover.
Turn margin to back of cover mitering corners.

③ Glue letters onto front.

② Glue fabric onto back of cover.

Notebook

1 cm

1 cm

Pencil Case

MATERIALS:

Navy and white checked quilted fabric, 22cm by 19cm. Navy felt, 32cm by 18cm. Zipper, 42cm long. Glue.

FINISHED SIZE: 8.5cm by 18cm.

CUTTING DIAGRAMS

Add no seam allowance unless otherwise indicated in parentheses.

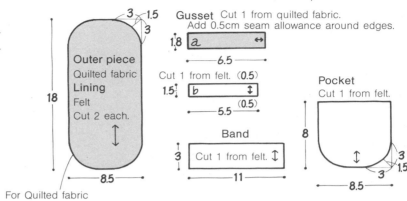

Outer piece
Quilted fabric
Lining
Felt
Cut 2 each.

For Quilted fabric

Gusset Cut 1 from quilted fabric.
Add 0.5cm seam allowance around edges.

a

Cut 1 from felt. (0.5)

b

Band
Cut 1 from felt.

Pocket
Cut 1 from felt.

DIRECTIONS

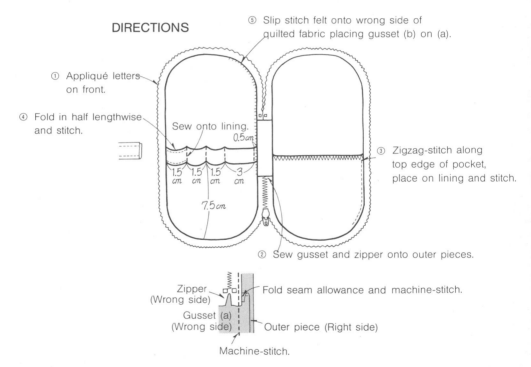

① Appliqué letters on front.

④ Fold in half lengthwise and stitch.

⑤ Slip stitch felt onto wrong side of quilted fabric placing gusset (b) on (a).

Sew onto lining.
0.5cm

③ Zigzag-stitch along top edge of pocket, place on lining and stitch.

1.5 cm 1.5 cm 1.5 cm 3 cm

7.5cm

② Sew gusset and zipper onto outer pieces.

Zipper (Wrong side)
Gusset (a) (Wrong side)
Fold seam allowance and machine-stitch.
Outer piece (Right side)
Machine-stitch.

Actual-size Appliqué Pattern

Glue appliqué pieces onto front and zigzag-stitch along edges of each letter.

Felt

PENCIL

Zigzag stitch

Continued from page 47.

Cup and Saucer

MATERIALS:

Floral print, 40cm square. Sheeting: White, 30cm by 20cm; dark brown, 8cm square. White felt, 25cm square.

Polyester fiberfill, 25g.

FINISHED SIZE: See diagrams.

CUTTING DIAGRAMS

Add 0.5cm for seam allowance.

Add no seam allowance to top edge of felt.

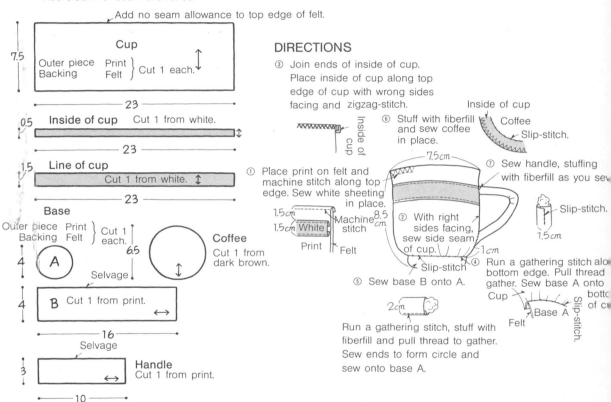

7.5

Cup
Outer piece   Print
Backing   Felt } Cut 1 each.

23

0.5

Inside of cup   Cut 1 from white.

23

1.5

Line of cup
Cut 1 from white.

23

Base
Outer piece   Print
Backing   Felt } Cut 1 each.

6.5

4   A   Selvage

Coffee
Cut 1 from dark brown.

4   B   Cut 1 from print.

16   Selvage

3   Handle   Cut 1 from print.

10

DIRECTIONS

③ Join ends of inside of cup. Place inside of cup along top edge of cup with wrong sides facing and zigzag-stitch.

⑥ Stuff with fiberfill and sew coffee in place.

Inside of cup   Coffee   Slip-stitch.

① Place print on felt and machine stitch along top edge. Sew white sheeting in place.

1.5cm   White   Machine stitch   8.5 cm   Print   Felt

7.5cm

⑦ Sew handle, stuffing with fiberfill as you sew

Slip-stitch.   1.5cm

② With right sides facing, sew side seam of cup.

Slip-stitch

1cm

④ Run a gathering stitch along bottom edge. Pull thread gather. Sew base A onto bottom of cup

⑤ Sew base B onto A.

2cm

Run a gathering stitch, stuff with fiberfill and pull thread to gather. Sew ends to form circle and sew onto base A.

Cup   Base A   Felt   Slip-stitch.

CUTTING DIAGRAMS FOR SAUCER

Add seam allowance indicated in parentheses.

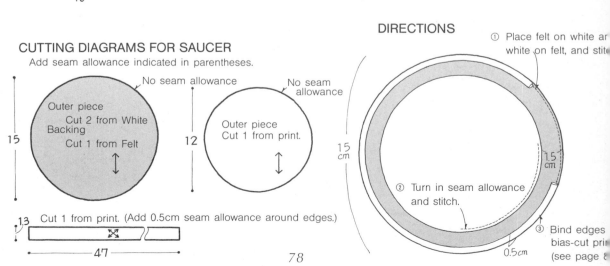

No seam allowance

Outer piece
Cut 2 from White
Backing
Cut 1 from Felt

15

No seam allowance

Outer piece
Cut 1 from print

12

1.3   Cut 1 from print. (Add 0.5cm seam allowance around edges.)

47

DIRECTIONS

① Place felt on white and white on felt, and stitch

15 cm

② Turn in seam allowance and stitch.

1.5 cm

③ Bind edges bias-cut print (see page 8

0.5cm

Cake and Plate
MATERIALS:
Sheeting: Beige, 36cm by 15cm; dark brown, 36cm by 12cm; white, 44cm by 22cm. Red Indian cotton fabric, 6cm square. Floral print, 50cm square. White felt, 40cm square. Polyester fiberfill, 25g.
FINISHED SIZE: See diagrams.

CUTTING DIAGRAMS FOR CAKE

Add 0.5cm for seam allowance.

2.5

7.5

Top and Bottom ↕

Outer piece
Backing

Cut 1 each from dark brown and beige.
Cut 2 from felt.

12.5

4.5 ↕

Cut 1 from red.

Side

5

A

Outer piece
Backing } Cut 1 each from beige and felt. ↕

35

B   Cut 1 from dark brown. ↕

1.5

35

DIRECTIONS

⑥ Run a gathering stitch along edge. Stuff with fiberfill and pull thread. Sew in place.

① Place dark brown on felt and stitch along edges. Stitch pieces for bottom in same manner.

Dark brown Machine-stitch.

④ With right sides facing, sew top, side and bottom pieces together leaving opening for turning. Turn to right side.

③ With right sides facing, sew side seam.

⑤ Stuff with fiberfill and slip-stitch opening closed.

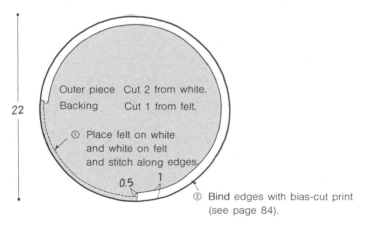

2.5 cm

Dark brown
Slip stitch.

1cm   12.5cm

2cm   A

5 cm   B

Beige   8cm Opening for turning

② Sew dark brown strip onto beige side piece.

1.5cm   Dark brown
Beige
Felt

CUTTING DIAGRAMS AND DIRECTIONS FOR PLATE

Add seam allowance indicated in parentheses.

22

Outer piece   Cut 2 from white.
Backing   Cut 1 from felt.

① Place felt on white and white on felt and stitch along edges.

0.5   1

② Bind edges with bias-cut print (see page 84).

(Add 0.5 cm seam allowance around edges.)

2   Cut 1 from print.   ✕

65

Continued from page 55.

Shoe Case and Cosmetic Case

MATERIALS:

Blue bonded cotton fabric: 66cm by 23cm for Shoe Case and 50cm by 20cm for Cosmetic Case. For Shoe Case: White felt, 9cm by 3cm. Six strand embroidery floss, No. 25 in blue. White cotton tape, 2.5cm by 27cm. Snap fasteners, 1cm in diameter, 2 pairs. For Cosmetic Case: White felt, 9cm by 3cm. Six strand embroidery floss in blue. Navy zipper, 28cm long.

FINISHED SIZE: Shoe Case, 21cm by 30cm. See diagram for Cosmetic Case.

CUTTING DIAGRAM AND DIRECTIONS FOR SHOE CASE

Add seam allowance indicated in parentheses.

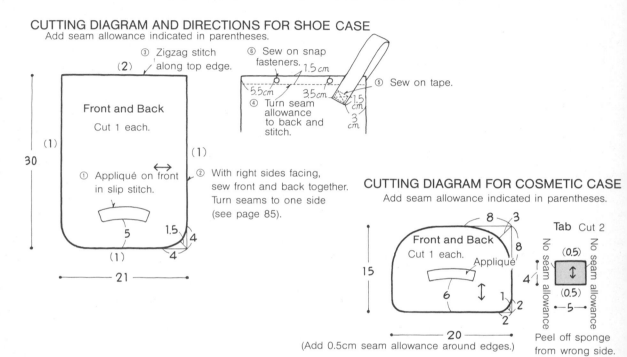

③ Zigzag stitch along top edge.

⑥ Sew on snap fasteners.

⑤ Sew on tape.

④ Turn seam allowance to back and stitch.

Front and Back
Cut 1 each.

① Appliqué on front in slip stitch.

② With right sides facing, sew front and back together. Turn seams to one side (see page 85).

CUTTING DIAGRAM FOR COSMETIC CASE

Add seam allowance indicated in parentheses.

Front and Back
Cut 1 each.

Appliqué

(Add 0.5cm seam allowance around edges.)

Tab Cut 2

No seam allowance

Peel off sponge from wrong side.

Gusset Cut 1.

Actual-size Pattern
Use 6 strands of floss.

Felt

Back stitch

DIRECTIONS

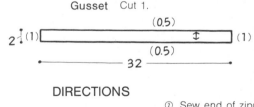

② Sew end of zipper gusset together with folded tab in between. Zipper.

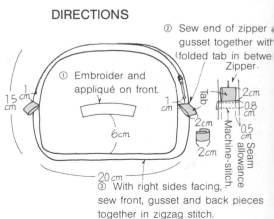

① Embroider and appliqué on front.

Tab

Machine-stitch.

③ With right sides facing, sew front, gusset and back pieces together in zigzag stitch.

Continued from page 59.

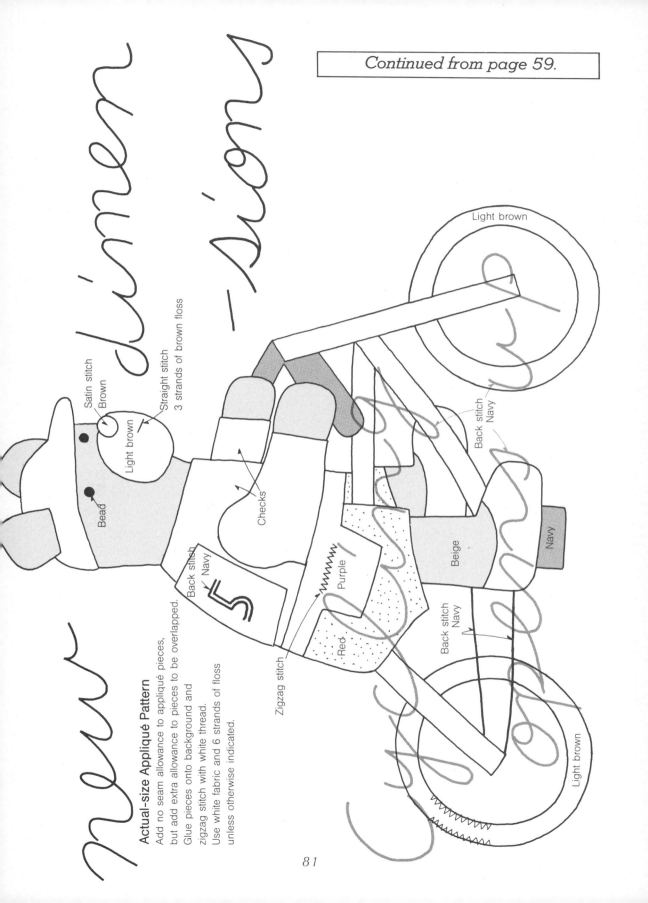

Actual-size Appliqué Pattern

Add no seam allowance to appliqué pieces,
but add extra allowance to pieces to be overlapped.
Glue pieces onto background and
zigzag stitch with white thread.
Use white fabric and 6 strands of floss
unless otherwise indicated.

Satin stitch
Brown

Light brown

Straight stitch
3 strands of brown floss

Light brown

Bead

Back stitch
Navy

Checks

Zigzag stitch

Back stitch
Navy

Purple

Red

Back stitch
Navy

Beige

Back stitch
Navy

Navy

Light brown

Light brown

81

Pochette, *shown on page 57.*

MATERIALS:
Mustard quilted fabric, 75cm by 28cm. Fabric for appliqué: Beige, white and gray sheeting, 6cm square each. Scrap of brown felt. Six strand embroidery floss, No. 25: Half skein of charcoal gray; small amount of light brown. Zipper, 25cm long. One pair of D-shaped buckles, 1.8cm in inner diameter. Glue.
FINISHED SIZE: See diagram.

CUTTING DIAGRAMS

Add 1cm for seam allowance.

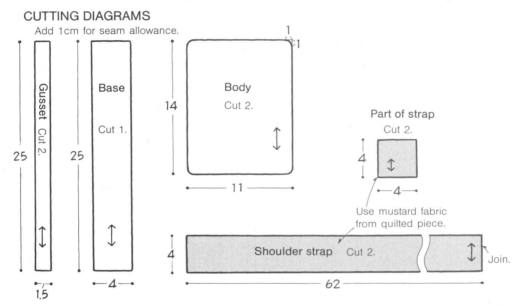

Gusset Cut 2. 25 — 1.5

Base Cut 1. 25 — 4

Body Cut 2. — 14 — 11 — 1, 1

Part of strap Cut 2. — 4 — 4
Use mustard fabric from quilted piece.

Shoulder strap Cut 2. — 4 — 62 — Join.

DIRECTIONS

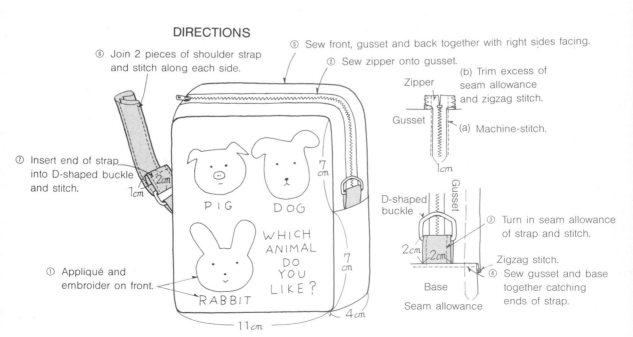

⑥ Join 2 pieces of shoulder strap and stitch along each side.

⑤ Sew front, gusset and back together with right sides facing.

② Sew zipper onto gusset.

⑦ Insert end of strap into D-shaped buckle and stitch.
2cm 1cm

① Appliqué and embroider on front.

PIG   DOG
WHICH ANIMAL DO YOU LIKE?
RABBIT

7cm   7cm   4cm
11cm

(b) Trim excess of seam allowance and zigzag stitch.
Zipper
Gusset
(a) Machine-stitch.
1cm

D-shaped buckle
Gusset
③ Turn in seam allowance of strap and stitch.
2cm   2cm
Zigzag stitch.
④ Sew gusset and base together catching ends of strap.
Base
Seam allowance

82

Actual-size Patterns

Use charcoal gray unless otherwise indicated.
Use 6 strands of floss for letters and 2 strands for other embroidery.

Beige

French knot

Light brown
Slip stitch.

Brown

Straight stitch

PIG

Gray

Satin stitch.

Straight stitch

DOG

Back stitch

White

Add no seam allowance
to appliqué pieces.
Glue pieces onto background
and zigzag stitch with
white sewing thread.

Fly stitch

RABBIT

**WHICH
ANIMAL
DO
YOU
LIKE ?**

Helpful Tips for Embroidery and Sewing

How to transfer design to fabric

Trace actual-size design onto tracing paper with a hard pencil. Place dressmaker's carbon paper between fabric and traced design and trace with a steel pen. Use carbon paper in color similar to embroidery thread.

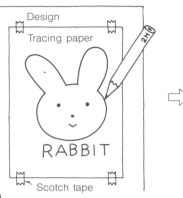

Design

Tracing paper

RABBIT

Scotch tape

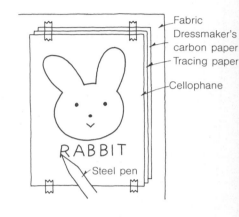

Fabric
Dressmaker's carbon paper
Tracing paper

Cellophane

RABBIT

Steel pen

How to alter scale of design

Divide design into squares. Prepare second paper with same number of sections according to the size desired. Copy design onto second paper.

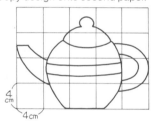

4 cm

4 cm

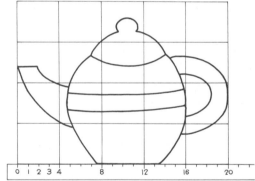

0 1 2 3 4   8   12   16   20

How to make bias tape

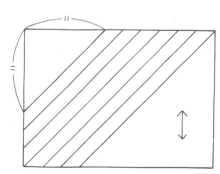

Cut strips on the bias along marked lines.

Then join ends of strips with right sides facing.

(Right side)

(Wrong side)

Trim off excess.

How to bind

Sew bias-cut strip along edge with right sides facing.

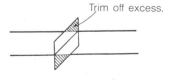

Machine-stitch.
Bias-cut strip (Wrong side)

(Right side)

Turn strip to back, turn in seam allowance and slip stitch.

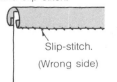

Slip-stitch.

(Wrong side)

How to finish seam allowance
Turn seam to one side.

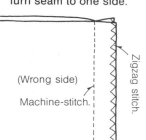

(Wrong side)

Machine-stitch.

Zigzag stitch.

Turn seam to one side and stitch along seam.

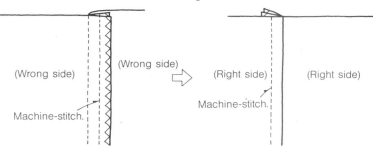

(Wrong side)

(Wrong side)

Machine-stitch.

(Right side)

(Right side)

Machine-stitch.

Basic Embroidery Stitches
Straight stitch

Running stitch

Back stitch

Outline stitch

French knot

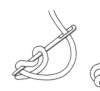

Fly stitch

Satin stitch

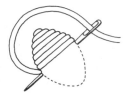

Couching

Cross stitch

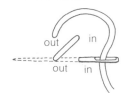

out in

out in

Bullion stitch

out

out

in

3 out

1 out

out

2
in

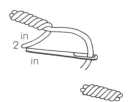

in

2

in

Slip stitch

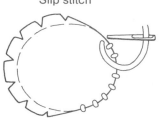